SEA LEVEL
RISE

SEA LEVEL

· A SLOW TSUNAMI ON AMERICA'S SHORES ·

RISE

Orrin H. Pilkey Jr. and Keith C. Pilkey

Duke University Press   *Durham and London*   2019

© 2019 Duke University Press
All rights reserved
Printed in the United States of America on acid-free paper ∞
Designed by Courtney Leigh Baker
Typeset in Minion Pro and Scala Sans by Copperline Book Services

Library of Congress Cataloging-in-Publication Data
Names: Pilkey, Orrin H., [date] author. | Pilkey, Keith C., [date] author.
Title: Sea level rise : a slow tsunami on America's shores / Orrin H. Pilkey Jr.
and Keith C. Pilkey.
Description: Durham : Duke University Press, 2019. | Includes
bibliographical references and index.
Identifiers: LCCN 2018061094 (print) | LCCN 2019005519 (ebook)
ISBN 9781478006374 (ebook)
ISBN 9781478005063 (hardcover : alk. paper)
ISBN 9781478005124 (pbk. : alk. paper)
Subjects: LCSH: Sea level—United States. | Environmental refugees—
United States. | Climatic changes—Social aspects—United States. | Climatic
changes—Health aspects—United States. | Climatic changes—Political
aspects—United States. | Climatic changes—Economic aspects—United
States. | Shore protection—United States. | National Flood Insurance
program (U.S.) | Flood insurance—United States.
Classification: LCC GC90.U5 (ebook) | LCC GC90.U5 P55 2019 (print) |
DDC 363.738/743—dc23
LC record available at https://lccn.loc.gov/2018061094

Cover art: King tide, Annapolis, Maryland, 2012. Courtesy of Amy
McGovern, Creative Commons.

· CONTENTS ·

· ACKNOWLEDGMENTS ·

When all was said and done, we learned that the impact of a rising sea on the shores of America is as wide-ranging and complex as are the shores themselves. And we discovered that the views of people living near the shore are also wide-ranging, spanning from doubters to believers and from let's-wait-and-see types to let's-do-something-about-it types.

The whole exercise of writing this book was an extraordinary adventure in nature on a very large scale. We watched from afar the interactions of the oceans versus the lands. We watched the huge mass of the oceans expanding, reflecting both a warming atmosphere and the melting of Earth's massive ice sheets. We learned that scientific opinion varies quite a bit also. Very few scientists doubt the validity of global climate change, but there is a wide range of viewpoints on the direction and magnitude of the earth's changes. For example, in the all-important estimate of sea level rise by the year 2100, estimates from credible scientists range from 1 foot to a rather wild 10 feet. And we became convinced that great changes are coming to America because of sea level rise, including the loss of two major cities, Miami and New Orleans.

We relied heavily on the research and the intuition of many scientists. Owen Mason, an expert on the ways of the Arctic coast, introduced us to Shishmaref and Kivalina, Alaska, and the serious problems they face in a rising sea. We gained particular understanding of the problems during two trips to these villages in the Arctic winter. Denali Commission Federal Co-chair Joel Neimeyer helped us understand the politics of the American Arctic. Stan Riggs led us on an Inner Banks, North Carolina, field trip, educating us on how roads raised to prevent flooding can exacerbate flooding. They do so by acting as dams during storm surges. University of Miami geologist Hal Wanless, with his constant pounding of the table about the dismal future of a flooding Miami and South Florida, was an inspiration for us because he continues to sound the alarm even though he's not widely appreciated. Al Hine and Tonya Clayton are from the state that will be most impacted by sea level rise, and they helped us understand Florida's situation. Geologist Alex Glass kept us abreast of the deteriorating condition of the world's ice sheets and Peter Haff helped keep us abreast of the exponentially expanding literature on sea level rise and provided us with references, ideas, and inspiration that opened new, adventurous avenues of thinking. Daughter/sister Diane Pilkey, a senior nurse consultant, helped us with the health chapter, the subject in which we were least competent. Geologist Bill Neal told us about the similarity of the fate of Newfoundland's small villages and the apparent future of small coastal villages in Alaska.

Whenever we pushed the wrong button on our computers, ever-helpful Andy Minnis and Katheryne Doughty, IT experts, saved us. Without going into all the details, there were many others who helped us. Among them were scientists Joe Kelley, Andrew Cooper, Duncan Heron, Andy Coburn, and Rob Young. Fred Dodson and Marcia Tuttle provided ideas and timely edits from the standpoint of ones who know relatively little about the topic.

Geological technician and able research assistant Norma Longo carried out many hours of media research on a number of topics and provided proofing, editing, and just about everything else leading to the production of the manuscript and getting it to the press. We are very grateful for her unwavering support and patience. We are also very grateful for the assistance of our editor, Gisela Fosado, and editorial associate, Alejandra Mejia, and the staff at Duke University Press for their guidance and assistance in completing this book.

We also want to express our gratitude to Claire Le Guern Lytle, who maintains and monitors the website Coastalcare.org, a gold mine of global coastal information from which we obtained photos and ideas for this book. The website is sponsored by our friends Olaf and Eva Guerrand-Hermes through the Santa Aguila Foundation, a nonprofit foundation dedicated to defending the beaches and shorelines of our delicate planet.

We are most grateful for the foundation's support of our recent books, including *The World's Beaches, Global Climate Change, The Last Beach, Lessons from the Sand, The Magic Dolphin,* and *Retreat from a Rising Sea.*

Financial support for the production of the book came from North Carolina state representative Pricey Harrison and from the Duke University Nicholas School of the Environment.

· PROLOGUE ·

When a tsunami strikes a developed shoreline, the damage can be severe and instantaneous. When the sea level rises, the damage will also be severe but will occur slowly, over a period of years and decades. There are some other obvious differences. Sea level rise is not caused by earthquakes or volcanoes and will not readily subside but will continue for decades and centuries, perhaps millennia. And unlike the relatively regionalized impact of tsunamis, sea level rise will impact all the world's ocean shorelines. To emphasize the threat to our coast, we've chosen to describe sea level rise as a slow-motion tsunami.

The rising sea may be the first truly worldwide catastrophe caused by global climate change. It will impact all seven continents and all the world's coastal cities from Los Angeles to New York, Rotterdam, Lagos, Mumbai, Shanghai, Tokyo, Honolulu, and many others. A few, like Miami and New Orleans, will disappear, as their geographical features guarantee that they ultimately cannot be defended against the rising waters.

Since all the world's port facilities are at the same elevation, the world's economy will be impacted as docks, warehouses, and freight yards the world over will need to be raised—a process that must be repeated for centuries to come as the waters continue to rise due to the melting of the world's land-based ice. Oceanside tourist facilities will require reconstruction, raising up, moving, and in most cases, ultimately, abandonment. Finally, many millions of people will become refugees fleeing from the rising sea, perhaps within this century. We refer to these throughout the book as *climate refugees* because the storms and the sea level rise that are forcing the exodus from low-lying coastal regions are affected by global climate change.

In this book, we do not concern ourselves with the mechanics and causes of global sea level rise. It's already under way. Instead, we start with the assumption that the scientific consensus concerning the magnitude of the sea level rise, with all of its pluses or minuses, is correct. We emphasize the impact of sea level rise in America only.

We accept that a 3-foot sea level rise by the year 2100 is a good possibility in the context of our current understanding of all the processes that are raising sea level. It is a certainty that the magnitude and direction of these processes (melting ice and expanding oceans) will change in coming decades. Most likely that change will lead to an increase in the rate of sea level rise. The current minimum expected sea level rise by 2100 is around 1 foot, and the maximum is 6 feet. Field observations indicate that increasing instability of the Greenland and Antarctic ice sheets has led to the prediction by some that it is possible but unlikely that sea level will rise by 10 feet by the century's end. The instability of the West Antarctic Ice Sheet was demonstrated in an extraordinary article published in the June 13, 2018, issue of *Nature*. The article, authored by 84 scientists from 44 institutions located in 14 countries, using data from 15 satellites plus field observations, concluded that the rate of melting of Antarctic ice has tripled since 2012. Of particular concern is the massive Thwaites Glacier, which appears to be destabilizing.

Late 2018 brought a flurry of conference results and studies, each strongly supporting the proposition that response to global climate change, including sea level rise, is critical and needs immediate action. Most important of these responses is the reduction of carbon dioxide emissions.

- On October 6, 2018, the United Nations Intergovernmental Panel on Climate Change (IPCC) released a stunning report, three years in the making. The panel suggested that global warming is on track, to a critical 1.5 degrees Celsius (2.7 degrees Fahrenheit) above preindustrial level temperatures. Within 30 years, global problems with drought, floods, fires, sea level rise, and massive migration will likely reach critical stages if this degree of temperature rise occurs.
- On November 20, 2018, the World Meteorological Congress reported that atmospheric concentrations of carbon dioxide reached a new high of 405.5 parts per million in 2017.

- On Black Friday in November 2018, the federal government released the fourth National Climate Assessment. The document was produced by 13 separate agencies and emphasizes "The United States is already suffering economic and public health damages from climate change-fueled wild fires, heat waves and floods and these damages will get worse if we don't take bold action to address it."
- On December 3, 2018, the National Oceanic and Atmospheric Agency issued the annual Arctic Report Card and noted the Arctic is warming at twice the rate of the rest of the globe.
- On December 5, 2018, the Global Carbon Project reported that global carbon dioxide emissions reached an all-time high in 2018. It is widely accepted that human-produced carbon dioxide underlies global climate change.
- On December 11, 2018, UN climate talks in Poland reinforced the Paris Climate Agreement. The U.S. representatives argued unapologetically that a "rapid retreat" from fossil fuel use was unrealistic because of potential damage to the economy, a position ignoring the catastrophic damage we face from climate change.

At about the same time, in an extraordinary development, the editorial boards of the *Florida Sun Sentinel*, the *Miami Herald*, and the *Palm Beach Post* have joined forces to raise awareness in South Florida about the huge threat that region faces from sea level rise—more than in any other state.

The Magnitude of the American Problem

For the sake of convenience, we focus on what could be expected from sea level rise up to 2100. *But it is very important to recognize that we are actually dealing with a 400-year (and more) problem.* Sea level rise is occurring because of warming and expanding ocean waters and the melting giant ice sheets add to the volume of ocean waters. The warming is caused mainly by the heat-trapping effect of greenhouse gasses, such as carbon dioxide (the main one) and methane. Humans are largely responsible for the increasing greenhouse gasses in the atmosphere through emissions from the burning of fossil fuels (coal, natural gas, and oil). The gasses trapped

TABLE P.1 The Length of American Shorelines

	Ocean shoreline miles	Tidal shoreline miles
American shorelines	12,400	88,600
The Lower 48	4,993	53,677
East Coast	2,069	28,673
Gulf Coast	1,631	17,141
Pacific Coast w/o Alaska	1,293	7,863
Alaska—Arctic and Pacific	6,640	33,904
Hawaii	750	1,052
Florida, both sides	1,350	8,426
Puerto Rico	311	700
Virgin Islands	117	175

SOURCE: National Oceanic and Atmospheric Administration (NOAA).

in the atmosphere today will cause melting of ice and rising of seas for centuries to come.

As is apparent in table P.1, the United States has plenty of shoreline that will be impacted by sea level rise. This country has 12,400 miles of shoreline facing the open ocean. Including bays, lagoons, and estuaries, the total number of miles of American shoreline affected by ocean tides comes to more than 88,600 miles.

Much of the 88,600 miles of American shoreline is developed to one degree or another—development that ranges from farms to small native Alaskan villages to major cities to miles and miles of beach cottage–lined shores to many miles of long rows of massive high-rise condominiums and hotels. The impacts along these miles of shoreline will be immense because of the expected long-term sea level rise.

Recently, Silicon Valley startups (e.g., Jupiter and Coastal Risk Consulting) have come on the scene, with plans to predict the risks that businesses and communities may face over coming decades from heavy rains, sea level rise, storm surges, and other climate change-related threats. A company called FM Global does the same for insurance companies. Some academic institutions, such as Western Carolina University, with its Program for the Study of Developed Shorelines, are entering the community risk-analysis business. The success of the commercial entities remains to be seen, but their establishment reflects the growing public recognition of the threats from and the reality of global climate change.

Of course, the coastal risk hazards are not evenly distributed around our coasts. A 2017 research report by Climate Central listed the top 25 cities in the United States and their populations at risk of flooding within the Federal Emergency Management Agency's (FEMA) 100-year coastal flood-plain. New York City has by far the largest population at risk (426,000). The extraordinary risk of Florida communities to flooding and sea level rise is indicated by the fact that out of the top 25 communities, only 5 are not in Florida (New York, Charleston, Virginia Beach, Norfolk, and Boston).

It is essential that Americans have a good grasp of what sea level rise has in store for us along our shorelines—politically, environmentally, and economically. We conclude that we are left with two choices. We can respond to sea level rise now in a peaceful, organized way, or we can respond later in crisis mode in reaction to flooding and storm disasters. If we plan now, we will save future generations from visiting shorelines with no beaches along a coast littered with rubble from destroyed seawalls, roads, and building foundations, looking like a battleground where the battle was lost.

. . .

This is the third book on sea level rise that Orrin Pilkey, the senior author, has written, with two of the three co-authored by Keith Pilkey. An earlier one (*The Rising Sea*, with coauthor Rob Young) was concerned with the mechanics of sea level rise, global warming skeptics, and the global impact. *Retreat from a Rising Sea*, by three Pilkeys (Orrin, Keith, and Linda), covers all aspects of sea level rise and the necessary retreat from the shore. The present book is concerned primarily with the impact of sea level rise on American shores.

· THE RISING SEA ·

Can't you see the climates changing?
Mother Earth is rearranging.

A rising sea laps at my feet
Prairies scorched by soaring heat

Shriveled crops cry out from thirst
Some say we ain't seen the worst

Critics claim it's nature's way
Man's impact has played no sway

Seas may swell and glaciers melt
Before the full effects are felt

Cities drown, shorelines retreat
Streams disappear in sweltering heat

Mother Earth cries in despair
Why don't people really care?

Skeptics claim it's all a myth
Wallowing in apathy and ignorant bliss

Denying facts, it's just a hoax
To calm the fears that truth evokes

But let's unite and take a stand
To face the facts and save the land

If not, what will the children say?
A generation or two away?

Will they cry out in despair?
Why didn't people really care?

—*Ronald D. Perkins*

FLEE THE SEA

· CLIMATE REFUGEES ·

"Sea-level rise is the defining issue of the century," declared a May 2018 editorial from Florida's *Sun Sentinel* newspaper. Ostensibly, the paper's editorial board was sounding the warning about the threat to South Florida, but this declaration applies to nearly all coastal communities in the United States.

As the sea continues to rise, coastal communities will lose territory, and they will lose people. The flow of sea-level-rise refugees is just a dribble now, with the major exception being Puerto Rico where, according to the U.S. Census Bureau, 130,000 people left the island in the year after Hurricane Maria, which struck in September 2017. The post-Maria exodus resulted in a 3.9 percent drop in Puerto Rico's population. A few souls have moved away from Miami and Miami Beach and a number of small towns on the Mississippi Delta, but the dribble is showing signs of acceleration as is the sea level rise.

Flooding from high tides, made higher by sea level rise, has forced the congregation of a Unitarian Church in Norfolk, Virginia, to abandon the church building and move to higher ground. A few Native Alaskans have moved from their tiny beachfront villages into towns because of the rising sea and the loss of protective sea ice along the Chukchi Sea and Arctic Ocean shores. It is likely that the rest will follow. A stretch of the South Nags Head, North Carolina, shoreline has beachfront houses that were third-row houses 25 years ago. The 4,800-ton Cape Hatteras, North Carolina, lighthouse, threatened by erosion, was moved back 2,900 feet from the eroding shoreline—the same distance from the sea that it was when it was built in 1870. After Hurricane Sandy in 2012, a number of houses along the low-elevation rims of Staten Island, New York, were purchased and removed to get them out of the way of future storms enhanced by sea level rise. In Bay Point, New Jersey, a small hamlet on the shores of Delaware Bay, 20 high-risk houses were purchased and demolished in 2018. Farmlands adjacent to the shores of the Delaware and Chesapeake bays and Pamlico and Albemarle sounds are increasingly impacted by salt water. On aptly named Washaway Beach in Washington State, shoreline retreat is very rapid, and as many as five artesian well water pipes can be seen protruding out of the water in a line offshore, each marking the site of a fallen, abandoned (and washed away) house. The U.S. Geological Survey noted that during the 2015–16 El Niño, shoreline erosion along the whole Pacific Coast between Mexico and Canada, and especially off Southern California was the worst it had been in 150 years. Much of the dune protection for low areas was lost. All these examples show a creeping loss of property is forcing people to move.

But there is another side to the story. In spite of widespread recognition of the continuing sea level rise, the abundant field evidence that it's occurring, and the dribble of refugees, we still seem to be thumbing our noses at it. Developers still construct beachfront houses along the Carolina and Florida shorelines. In North Carolina, officials allowed new rows of beach houses to be built next to newly nourished but rapidly eroding beaches. Almost without exception, the new houses are McMansions— three-story, multi-bedroom rental palaces. Unquestionably, the bigger the buildings, the fewer the options for response to the rising sea. In Miami, certainly the most threatened city in America, two high-rise construction projects costing more than a billion dollars each are under way. In Waveland, Mississippi, beachfront lots occupied by houses destroyed first in Hurricane Camille (1969), and then in Hurricane Katrina (2005), are once again, for the third time, occupied by new homes. Miles and miles of

FIGURE 1.1 A 2007 photo of houses on North Topsail Island, North Carolina, that were later demolished to make a useable recreational beach. The cost of building removal was about $30,000 each. Now, the next row of houses is threatened and will soon be removed or demolished. Ironically, most of this community was in a COBRA zone wherein no federal insurance or federal funding for infrastructure is available. We believe that North Topsail Beach is the most unsuitable barrier island segment for development on the East Coast and the most endangered by sea level rise. *Photo courtesy of Norma Longo.*

beach-destroying seawalls in the United States are under construction or repair in an attempt to postpone the date when the occupants must flee.

But in time, the dribble will change to a rush. Just imagine, four million Americans, the majority of them Floridians, forming a stream of refugees moving to higher ground. They will not be the bedraggled families carrying their few possessions on their backs as we have seen in countless photos of people fleeing wars and ethnic cleansing, most recently in Myanmar and Syria. Instead, they will be well-off Americans driving to a new life in their cars, with moving trucks behind, carrying a lifetime of memories and possessions.

There is no conceivable scenario by which we can stay near today's shore-

FIGURE 1.2 The home site in Waveland, Mississippi, of the parents and grandparents of the authors. The house at elevation 13 feet was flooded in Hurricane Camille (1969) and completely destroyed by Hurricane Katrina (2005). These events inspired the writing of this book.

line, with its beautiful sea view, as the sea level rises three feet or more by 2100. In order to do so, we would have to hold the shoreline in place using massive seawalls that would grow bigger and higher every decade, which can't be done while simultaneously preserving the all-important beach. Retreat is the only answer to a gloomy future of beachless beach communities lined by the rubble of destroyed buildings or by massive seawalls.

Getting individual beach cottage owners to move back to safer ground will be the easy part—everyone has an incentive to get out of the water! Increasingly, those living beside the beach will suffer from storm fatigue as they repeatedly rebuild after floods or hurricanes. The hardest part will be dealing with towns and cities, as there will be large numbers of people simultaneously affected, all asking for taxpayer support, first to hold off the sea level rise, next to help refugees start a new life, and finally to help the towns that will receive the refugees.

It won't happen all at once. This refugee crisis will play out over several decades, perhaps starting in earnest 50 years from now, as coastal dwellers with a fine view of the sea that has risen three feet find they must leave to survive. Or they must leave because their neighbors and friends have already gone. Or they must leave because nuisance flooding frequently floods and closes the roads that lead to schools, churches, hospitals, and businesses. The trickle of refugees will swell, propelled by the intensified storm surges and winds of hurricanes, likely more powerful as a result of both sea level rise and ocean warming.

If the sea rises three feet, the number of moving Americans may exceed 4 million, and if the rise is six feet by 2100, the number of refugees may exceed 13 million, according to a study by Mathew Hauer, head of the Applied Demography Program at the University of Georgia, and his associates. This estimate of American refugees will very likely prove to be conservative, especially in decades beyond 2100.

But it's not just a matter of future inundation of coastal properties. Low-lying lands, especially along the Gulf and Atlantic coastal plains, already are plagued by periodic nuisance flooding—otherwise known as *tidal flooding* or *sunny-day flooding*—defined as temporary flooding at high tides that causes public inconvenience. This flooding is also termed *chronic flooding* or *chronic inundation*, discussed further in chapter 3. The National Oceanic and Atmospheric Administration (NOAA) says that in 1950 nuisance flooding was infrequent and usually brought about by winds, but by 2010, nuisance flooding began to occur frequently, on sunny days, along American shorelines. The reason? Sea level is higher, and therefore normal high tides are higher than ever. Current nuisance flooding hot spots include Annapolis, Maryland, Norfolk, Virginia, and Miami Beach, Florida. How soon will their residents become refugees?

Refugee numbers will increase as repetitive nuisance flooding and ever more powerful hurricanes and storm surges penetrate their homes or neighborhoods well above the still-water or high-tide level of the sea. In addition, saltwater pollution of drinking water or intrusion of salt water into farmlands will create refugees as coastal living conditions deteriorate.

Even individual coastal developments at relatively safe elevations may have to be abandoned if surrounding streets at lower elevations become flooded and prevent access and egress. During Hurricane Harvey (2017), the Texas Medical Center was well equipped with generators and flood-walls, but the hospital itself was inaccessible because surrounding streets were flooded. Nobody could come and go (without a boat or helicopter).

American Refugee Numbers

Most American climate refugees will remain in this country and probably flee to nearby cities. The very first climate refugees in America must have been Native Americans who had to flee the rising sea that followed the end of the last glaciation. Relics from Native American settlements have been found under water on all our continental shelves. The rate of the retreat of shorelines over the last 10,000 years when the continental shelves off the Americas were inhabited must have varied widely depending on the slope of the now-submerged land. For example, the continental shelf of the Northeastern Gulf of Mexico, off the Apalachicola Delta, is extremely flat. Native Americans living alongside the shore at that time may have had to move their encampments every few months or so to escape the sea!

W. B. Cronin, in his book entitled *The Disappearing Islands of Chesapeake Bay*, notes that more than 500 islands have disappeared from Chesapeake Bay since European settlers arrived here. The presence of Native American relics on the remaining small island beaches indicates that the early Americans likely encamped there and on the now-disappeared islands and must have abandoned sites as islands disappeared beneath them. European settlers later inhabited the remaining islands, and their descendants were forced to leave as the islands eroded.

The first "official" American sea-level-rise refugees in modern history are the approximately 60 people who make up the Native American village of Isle de Jean Charles, Louisiana, on an island in the Mississippi River Delta. Its inhabitants, who used to number 400, are mainly from the Biloxi-Chitimacha-Choctaw tribe. The village is in trouble because the island is eroding and has been reduced in area by 90 percent. In 2016, the tribe was awarded $48 million for resettlement to higher ground. This amounts to a cost of $800,000 per inhabitant, a cost that is not likely to be repeated on a national basis in other communities that must eventually move. The hope is that this village will be reconstituted to its former population of 400 inhabitants once it has been moved. Unfortunately, the plan has a hitch, discussed in chapter 2.

In the Hauer study, the numbers of people at risk, mentioned in table 1.1, are rough approximations because:

- The future measures (such as nourished beaches, seawalls, and levees) that communities will take to hold back sea level rise and to reduce storm surge impacts are unknown. Such measures will reduce the number of refugees, but only temporarily.
- The refugee numbers may increase as shorelines retreat in the immediate future at rates that are difficult to predict, which would increase the impact of tidal flooding and storm surges.
- Some communities will be abandoned because, even though the communities themselves are not flooded, rising waters flood surrounding low-lying roads and crucial infrastructure, such as wastewater and drinking water plants.
- Some communities will be abandoned before many of the homes are seriously threatened because, as the more-threatened population flees, the underpinnings of what makes communities livable will be threatened—friends and neighbors leave, along with schools, businesses, and churches, and the tax base that supports local government shrinks.

Global Sea-Level-Rise Refugees

A 2017 Cornell University research report in *Science Daily* stated this: "In the year 2100, two billion people—about one-fifth of the world's population—could become climate-change refugees due to rising ocean levels. Those who once lived on coastlines will face displacement and resettlement bottlenecks as they seek habitable places inland."

Worldwide, arguably the first modern-day climate refugees are the hundreds of thousands of Indian and Bangladeshi people who have already fled their Ganges River Delta homeland. They leave behind infertile agricultural land (contaminated with salt), frequent floods, the constant threat of devastating typhoons, polluted drinking water, and failed small dikes, all amplified by a rising sea. Some scientists predict that by 2050, 17

TABLE 1.1 Possible Numbers of Climate Refugees

County	State	People at risk, 3-foot rise	People at risk, 6-foot rise
Miami-Dade	FL	231,336	1,967,018
Jefferson	LA	196,995	289,706
San Mateo	CA	190,189	249,020
Ocean	NJ	86,430	176,360
Charleston	SC	81,943	195,698
Suffolk	NY	69,625	145,291
Jefferson	TX	57,070	108,965
Virginia Beach	VA	46,017	143,226
Glynn	GA	23,795	48,127
Suffolk (Boston)	MA	19,735	168,000
Mobile	AL	17,017	27,799
Clatsop	OR	5,183	7,815

A listing of the possible number of refugees from various U.S. counties, for 3-foot and 6-foot sea level rises by 2100. These numbers are probably low because they reflect direct flooding and not the serious disruption of the entire county.

SOURCE: Adapted from Ghose, 2016.

percent of the Ganges Delta will have been lost due to erosion and flooding, which could eventually lead to 20 million sea-level refugees.

Estimates of the numbers of climate refugees from around the world in *noncoastal areas* are also horrific. According to a 2009 report, the Environmental Justice Foundation (EJF) noted that in just 10 or 20 years the number of refugees could number 10 or 20 million. Many of these will be driven out by climate change in the sub-Saharan area, and many will try to cross the Mediterranean Sea to Europe. Whole regions of the world may be rendered uninhabitable as we approach temperatures that are incompatible with human existence. The EJF urges governments to provide new legal frameworks to protect climate refugees and to take strong steps to reduce greenhouse gas emissions. New Zealand is currently considering creating a visa category to help Pacific refugees, mostly from atolls.

The United Nations has pointed out that climate refugees may actually lead to even more refugees. This is because in the competition for limited water, grazing land, and food, shortages could occur that will drive the native citizenry away or produce armed conflict, causing them to flee. Arguably, we have already witnessed climate-based conflicts in Syria and Sudan.

The American Hot Spots

NOAA has identified three sea-level-rise hot spots in the United States: the Inner Banks (the northeastern corner of North Carolina), the Mississippi Delta, and South Florida.

The Inner Banks Hot Spot

This is the lowermost coastal plain region surrounding Pamlico and Albemarle sounds, behind North Carolina's Outer Banks, and includes the corner of southeastern Virginia. The region is mostly very low-lying, often swampy (including parts of the Great Dismal Swamp), with a number of small towns, wildlife refuges, national forests, and superfarms that have closely spaced drainage canals.

The entire North Carolina Inner Banks population of around 280,000 souls likely will be forced to leave their homes, with even a two-foot sea level rise. Not only will such a rise inundate many towns, it will flood many access roads. And those roads that don't flood may actually contribute to flooding elsewhere. Geologist Stan Riggs of East Carolina University has observed that roads in the area, some already elevated three or four feet to keep traffic out of the water, act as dams that cause storm surges and river floods to inundate areas that otherwise would have remained dry. These road dams add considerably to the flood risk of many towns in the Inner Banks and were responsible for much damage in the floods from Hurricane Floyd (1999).

The Union of Concerned Scientists lists communities within the North Carolina Inner Banks that are susceptible to flooding. These are Alligator, Columbia, Croatan, East Lake, Fairfield, Fruitville, Gum Neck, Lake Landing, Lake Mattamuskeet, Pamlico, Swan Quarter, Cedar Island, Crawford, Harkers Island, Kinnakeet, Marshallberg, Salem, Scuppernong, Sea Level, Stacy, and Shiloh.

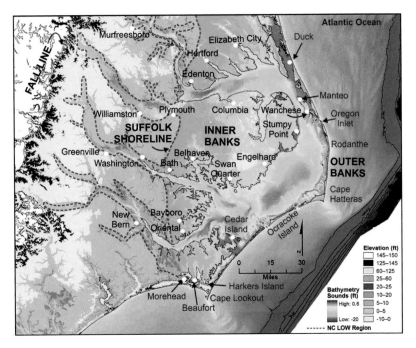

FIGURE 1.3 Map showing the aerial extent of North Carolina's Inner Banks, showing a number of towns within the Inner Banks that are at considerable risk from flooding from storm surges, nuisance flooding, and sea level rise. The low-elevation Inner Banks create a problem for people trying to evacuate from the Outer Banks. *Map courtesy of Stanley Riggs, NC LOW.*

The Florida Hot Spot

South Floridians streaming to the north will make up many of the American climate refugees. According to Hauer, 6 million or more refugees will flee to the north from Florida in a six-foot sea level rise. This may be a low estimate, because the highly porous and permeable limestone that underlies large parts of South Florida, particularly Miami and Ft. Lauderdale, creates a great hazard. Here, any seawalls, dikes, and levees that may be constructed will have little effect on sea-level-rise flooding. According to Harvard instructor Jesse Keenan, people are already moving away from areas that suffer from tidal flooding in both Miami Beach and Miami. With sea level rise, it is best to be ahead of the wave because if you wait too long your property values may collapse.

America has previously experienced migrations of large numbers of citizens within the nation's boundaries. During the nineteenth century, many thousands of Americans, a lot of them in covered wagons, streamed westward in search of land and wealth, forcibly removing Native Americans from their land in the process. As historian David McCullough described a later migration, "A shift of population that had begun during the war [World War II] grew to surprising proportions over the next thirty-odd years, becoming one of the nation's biggest migrations. Millions of people, black and white and mostly poor, left the rural South for the big cities of the North and West." These migrants, like the westward moving pioneers of the eighteenth and nineteenth centuries, also sought a better life. African Americans, in particular, fled the "Jim Crow" South for the less blatantly racist North.

But the future streams of climate refugees who will flee their homes, displaced by the rising sea, are not likely to be largely destitute. A large number of people probably will be fleeing at the same time, from the same place, vastly complicating the whole process of leaving home and resettling in some distant locality.

We predict that because of sea level rise, Florida will decide to protect buildings instead of beaches and will construct seawalls along many of the recreational beaches. As a consequence, their beaches will be gone by the turn of the century. When the beaches disappear, so will most of the tourists. The hotel/condominium usage will plummet, and the number of people employed in tourist-related positions will also decrease, adding to the refugee stream.

Where will all these millions of South Florida climate refugees end up? Since we're anticipating 50 years and more before the rush begins, and because economic opportunities probably will govern the choice of refuge, it is difficult to gauge the future. Will Tallahassee, safely located in North Florida well above any projected sea level rises, become a bigger city? Or will future cities be constructed along the Lake Wales Ridge, the sandy ridge that runs for 150 miles down the middle of the state? This ridge consists of remnants of ancient islands from a time when seas were much higher in the past. Perhaps we will witness the construction of entirely new cities to handle the coastal population disruption. We could use this opportunity to correct some of the failures of twentieth-century

U.S. cities with their inefficient public transportation and car-centric construction in favor of greener, pedestrian-friendly cities.

Hauer warns that "population surges could put insurmountable stress on local utilities, like water resources, if landlocked counties don't start preparing soon for global-warming-related migrants." He adds that "too much of the climate-change-preparedness debate has centered around preparing coastal communities for rising ocean levels—and not enough attention has been paid to preparing landlocked cities for new residents." In fact, to our knowledge, with the possible exception of LA SAFE, discussed later, serious attention is not being paid anywhere in the United States to the possibility of a future flood of refugees.

The Delta Hot Spot

River deltas are particularly desirable sites for development because of rich agricultural land, easy access to the sea for fishing and aquaculture, and economy-building port facilities. As a consequence, deltas all over the world have been highly impacted by humans. Besides the massive Mississippi Delta, the United States has several smaller marine deltas, such as the Colville and Yukon/Kuskokwim deltas in Alaska and the Sacramento/San Joaquin Delta in California.

Several factors combine to make deltas particularly susceptible to sea level rise. For instance, the amount of sediment contributed to the deltas by the rivers has been significantly reduced, sometimes to the point of extinction, by upstream sand-trapping dams. Water use by humans, such as irrigation, reduces the volume of water in rivers, thereby reducing the sediment load flowing toward deltas. And on the delta itself, levees, dikes, and navigation channels direct the river flow (and sediment load) offshore, preventing the natural spread of sediment that would normally build up the delta surface. Finally, extraction of oil, gas, and water causes sediment compaction (subsidence) or a sinking of the land and can significantly increase the rate of local sea level rise.

The one-two punch of rising seas and an expected increase in intensity of storms due to climate change arrives at a time when river deltas are already in a state of degradation. All modern river deltas are at low elevation, so as a consequence, rising seas will rapidly increase coastal flooding and erosion, reduce the size of wetland areas, and increase salinization of

farmland and groundwater. And low-lying deltas are particularly susceptible to hurricanes, as they tend to protrude out into the sea.

All of these delta generalizations are true for the 12,000-square-kilometer
Mississippi River Delta, 10,000 square kilometers of which are estuarine
wetlands. In 2005, two major hurricanes, Katrina and Rita, came ashore
on the Mississippi Delta, causing a total of 2,000 deaths. But the greatest
global delta disaster of all was caused by the Bhola hurricane in 1970 that
killed an estimated 500,000 people on the Ganges Delta in Bangladesh.

Because of the loss of sediment from upstream dams, the natural and
induced subsidence, and the rising sea level, most of the Mississippi Delta
will drown by 2100, according to a 2009 study by Louisiana State University professors Michael Blum and Harry Roberts. Most of the roughly two
million people who live on the Delta will become sea-level-rise refugees.

But residents of the Mississippi Delta show a degree of recognition of
the rising sea that other U.S. hot spots do not. Annie Snider in *Politico*
writes:

> With the help of $92.6 million in federal grant money, Louisiana's
> Office of Community Development has launched a first-of-its-kind
> effort (known as LA SAFE) to help communities across the state
> prepare for the tumult to come. Rising waters and escalating flood
> insurance rates will drive thousands of families farther inland, the
> state predicts, leaving behind homes they've known for genera-
> tions and places that have fundamentally shaped their identities.
> But those refugees aren't the only ones who will experience change.
> Communities like Houma will experience their own jarring transi-
> tion as they receive an influx of waterlogged neighbors. Houma sits
> high enough that it's less likely to drown in a hurricane. The idea of
> LA SAFE is to set an example for other communities that face the
> same sea level rise threats.

Not all delta dwellers are on board. Timothy Kerner, mayor of the
town of Jean Lafitte, population 7,000, proposes to build a $1 billion levee
around his tiny town and says, "No matter what, we're not leaving."

Appendix A is a list of the world's largest and most populous deltas
with estimates of the likely number of their refugees within the next 50
years.

Retreat as a Golden Opportunity

Clearly, no community or government entity at any level is considering the problem of having to accommodate thousands, even millions, of refugees fleeing their flooded homes. This flood of Americans seeking new homes, jobs, and schools for the kids is an absolute certainty in coming decades. It also represents an opportunity, a chance to do it right, to learn from the past and build with good construction practices at safe elevations. For example, the participants of the New York buyout program that purchased flooded low-elevation homes after Hurricane Sandy viewed the buyouts as a new lease on life. They moved to safer locations and no longer had to return, poststorm, to flooded houses. They felt safe and secure again in their new locations.

If we view the move in a positive way, it can be an unparalleled opportunity for new and revitalized industry and new modern communities; but it won't work if we wait until the wolf is at the door. Ideally the march of refugees should start shortly as the future sea level rise becomes more and more obvious and undeniable, as coastal house prices tank, and as nuisance flooding becomes a daily occurrence. Chances are, however, that many of the moves will come as a result of a superstorm, or a season of many storms, or year after year of damaging storms, at which point the stream of refugees will be an all-at-once regional event.

We are clearly facing an uncertain future forced on us by global climate change and particularly by a rising sea. Yet, instead of recognizing the urgency of our plight and the need for immediate long-term planning and action, we remain in a cocoon of questioning and uncertainty aided by some of our leaders who stare at the evidence of climate change and unstoppable sea level rise and yet deny it.

THE END OF THE
INUPIAT WAY OF LIFE

The rising sea is a dark cloud on the horizon for many coastal dwellers. But there is a particularly black cloud for some American communities that are on the brink of extinction. The Inupiat Eskimo people, who live primarily above the Arctic Circle, and other Alaskan native groups (including Inuits, Yupiks, Aleuts, and others) living in some 34 coastal villages in Alaska, are teetering on the edge of disaster caused by shoreline retreat brought on by a warming climate and a rising sea. These are the last near-subsistence societies in America.

Unalakleet, Shaktoolik, Golovin, Teller, Shishmaref, Deering, Selawik, Kivalina, and Barrow (now called Utqiaġbvik) are loosely considered Inupiat villages. Two of these small villages, Shishmaref and Kivalina, are situated on narrow barrier islands lining the Chukchi Sea shoreline of Alaska. Shishmaref lies just below the Arctic Circle and Kivalina just above the Arctic Circle. While Shishmaref and Kivalina are among the most seriously threatened of these native villages, twelve other villages are actively considering relocation. A few are along riverbanks (e.g., Newtok), but most lie along the shores of the Chukchi Sea, the Norton Sound

FIGURE 2.1 A Google Earth image of Shishmaref, Alaska, an Inupiat Eskimo village of 580 souls bordering the Chukchi Sea. Due to sea level rise, melting permafrost, and later time of formation of protective sea ice, erosion threatens the village. In the next few years, they must move to the mainland. *Photo source: Google Earth.*

within the Bering Sea, and along the Arctic Ocean shoreline south to the Canada-U.S. border.

The coastal native villages in the American Arctic represent essentially the only Native American societies not swallowed up by the rest of American society. They are not relegated to reservations, and they live at least partly the way their ancestors lived.

Many Alaskan Inupiat are also found in nearby towns, such as Nome and Kotzebue, but those folks no longer rely on the subsistence lifestyle. Populations in the native villages (Shishmaref, 580 people, and Kivalina, 400 people) are small by necessity. A subsistence living is impossible with larger populations because food sources would quickly become exhausted. Esau Sinnok, an Inupiat resident of Shishmaref, described the food sources to National Public Radio (NPR): "It's a community that relies on hunting and fishing," he said. "A majority of our diet comes from the land and the sea. We hunt for caribou, moose, musk ox, bearded seal, walrus, and gather traditional berries like the cloud berry, blueberries, blackberries." Bowhead whales, a traditional food for Inupiat living farther north along the Arctic Ocean shore (the Beaufort Sea), are not a reliable staple for Shishmaref inhabitants, but whale meat is often obtained by trading with whaling villages across the region.

Villagers do get much of their food by fishing and hunting, as Sinnok said, but they also enjoy such modern conveniences as satellite television, mobile phones, and game systems. The villages are not without other touches of modern society, as well, with electricity generators, rudimentary sewage service (honey buckets), and landfills. Plastic buckets serve as toilets that are emptied into larger containers, with the waste carried away in still larger buckets on trailers pulled by snowmobiles or four-wheelers.

The plight of Shishmaref has received, by far, the most publicity, mainly because it is easily accessible by scheduled air service, using a well-maintained hardened airstrip kept clear by a system of tractor-driven trailers (snowmobiles in the winter). This allows opportunities for air travel to and from larger towns, such as Kotzebue. Over the years, dozens of photographers, reporters, TV crews, documentary filmmakers, and a handful of scientists have visited the villages, taking advantage of a welcoming public relations campaign. The international media are particularly drawn to Shishmaref, with special TV programs originating from French, British, Japanese, and Dutch outlets.

A favorite scene for media visitors is a small deteriorating house that fell onto the beach in 2005 at the west end of town. It has remained purposely untouched more than a dozen years later and is perhaps one of the world's most photographed fallen houses, an iconic figure reproduced even in technical journals, such as *Science*. Only a few small houses have been moved back as erosion has proceeded, and new development has continued landward of the eroding frozen frontal dune.

Just like on the beaches of the lower forty-eight, sea level rise is a major cause of the shoreline retreat in the Arctic. But the Native Alaskans are faced with two other natural processes unique to high latitudes. One is the diminishing sea ice. A decade or two ago, sea ice formed at the Chukchi

FIGURE 2.2 This house in Shishmaref fell onto the beach during a storm in 2005. Since then, it has been maintained in place to provide visitors a highly visible example of the erosion problem. This may be among the most photographed failed beachfront buildings ever.

Sea shoreline starting in early September. Off Kivalina, the sea ice used to be 12 feet thick. Now it is just 4 feet thick.

Usually, the sea ice silences the surf zone and provides protection against shoreline erosion. Now, due to global warming, the protective sea-ice cover often doesn't freeze the surf zone until late November, so the storms of October and early November can directly attack the beaches and cause more shoreline retreat. In 2018, the Utqiaġvik surf zone in the Arctic Ocean did not ice over until January 1.

Other Arctic areas are dealing with similar situations. One impact in Greenland has been a recent increase in the number of dog sled teams. They are proving superior to snowmobiles for travel because the dogs can detect thin ice and guide the sleds around it.

The thinning ice has brought an element of modern living to Shishmaref. One resident bought a drone, which he claims is used mainly to distinguish weak ice from strong ice to make winter travel for hunting

FIGURE 2.3 Shishmaref, Alaska, during the winter, with a frozen surf zone. The sea ice shown here forms later in the year than in previous times so that late fall and early winter storms now hit the shoreline. The Chukchi Sea shoreline is to the right, and the lagoon shoreline is to the left. *Photo by ©Paul Andrew Lawrence/ www.Paulcolors.com.*

by snowmobile safer. The resident claims the drone will not be used to locate seals, a process that normally requires extensive searching by boat, dog sled, or snowmobile. Hopefully, the introduction of drone technology will not result in an overwhelming depletion of the seal population. Elsewhere, such as on the Grand Banks off Nova Scotia and Newfoundland, technological advances in fishing techniques with larger vessels and the use of echo sounders resulted in overfishing and eventual loss of the entire cod fishery.

In a larger picture, the loss of sea ice is not entirely bad. Climate change has made a reality of the once-fabled Northwest Passage. The expected increase in trans-Arctic shipping on the much-shortened route between East Asia and Europe will provide an economic boom, particularly for Canada, and perhaps a small part of this increased commerce/shipping could benefit indigenous populations.

FIGURE 2.4 Here is a mid-winter scene showing the frozen surf zone along a barrier island on the North Slope of Alaska. Obviously when the sea is iced over, the wave erosion problem doesn't exist, although onshore winds may push ice ashore and cause some erosion as it has here. The problem is that sea ice is forming progressively later in the fall and the beaches are subjected to October–November storm waves. *Photo courtesy of Andrew Short.*

The second unique process creating a problem is the melting of permafrost. Both Shishmaref and Kivalina lie in the zone of permafrost, which by definition is ground (including beaches) that is "permanently" frozen. But now, with the globally warming atmosphere, the beach sand remains largely unfrozen even during the winter. This means the beach, no longer protected by a combination of hard frozen sand and seasonal sea ice, is extremely vulnerable to attack by waves, especially the fall storm waves. Both islands have shrunk in size over the last few decades because of erosion on all sides.

To combat erosion, Shishmaref has seawalls—constructed in different designs in different places at different times—that may temporarily hold the shoreline in place. Prior to wall construction, the community first used discarded 55-gallon drums as protection along the shore, and in desperation in the midst of a storm, people discarded old dog sleds and snow-

FIGURE 2.5 This is one of the early seawalls in front of Shishmaref on Sarichef Island in Alaska. The wall of interlocking concrete blocks was ill-designed and, after construction, sand simply flowed out from beneath it, and the wall collapsed.

mobiles, abandoned furniture, and a variety of other old household items onto the beach. Kivalina also had a variety of seawalls, all of which failed.

Both communities have had ill-designed seawalls, typically of interlaced cement blocks, that failed in the first storm, days after construction was completed. In an attempt to protect the village of Kivalina, the U.S. Army Corps of Engineers (USACE) constructed a substantial seawall that is uniform along the entire ocean-facing shoreline. Portions of Shishmaref, especially on its north end, also have a substantial rock seawall. Both of these constructions will give the communities a little breathing room to plan their move.

Moving Back

Seawall or no seawall, both villages desperately need to be relocated, and the inhabitants have indeed voted to move. In contrast, the village of Shaktoolik (population 258), further south along the Bering Sea shore, has decided to stay put and try to defend their houses. FEMA has been approached for assistance, but the agency says federal disaster relief isn't meant to deal with the gradual impacts of climate change, like thawing permafrost and erosion. Ironically, if a giant storm destroyed a village, FEMA could help, but no such "luck."

The cost of moving the 580 residents of Shishmaref to the mainland has been estimated by USACE to be $140 million, which comes to nearly $250,000 per individual inhabitant. A normal cost-benefit ratio calculated for such a project would never fly. The high cost of moving the villages to safety from the rising seas is due primarily to the expensive costs of construction in melting permafrost terrain, as their new locations would also be on permafrost that, as it melts, causes buildings to sink into the soil.

Moving the village of Isle de Jean Charles, Louisiana, mentioned in chapter 1, will be an easier task than moving Shishmaref or Kivalina for several reasons: there is no permafrost, a road leads off the island, and the locals do not lead a subsistence lifestyle. But the cost estimates for moving Isle de Jean Charles are $800,000 per current inhabitant, much more than moving the individuals of Shishmaref. The assumption (and hope) is that hundreds of previous residents of Isle de Jean Charles may move back to the safety of a new town and the tribe will be reconstituted. The state plans to relocate the residents to a sugar cane farm outside of Houma, Louisiana.

Members of the Quinault Indian Nation living along the shores of the Olympic Peninsula in Washington State plan to relocate to nearby higher ground because of tidal flooding and shoreline erosion problems. Approximately 700 people from the village of Taholah are planning to move at an expected cost of $60 million. The move will include the police station, the courthouse, the post office, and most village homes.

So why preserve the Alaskan villages of a few hundred people, each at such high cost? What is their value to those of us who don't live in the Arctic and who have never been near the Arctic Circle?

Their value lies in the fact that the "threatened 34" are the only near-subsistence villages remaining in the United States. For this reason, preservation of the villages is preservation of a way of life. To ignore their serious plight, one caused by our own heavy consumption of fossil fuels and is of no fault of their own, is to continue with the same contempt shown indigenous American people throughout U.S. history.

Why don't the Native Alaskans simply move to the nearby cities (Nome, population 3,800; Kotzebue, population 3,200; or Anchorage, population 300,000)? Fifty-one percent of the population of Nome is already made up of Native Alaskans, and 71 percent of Kotzebue people are of native descent. These relocated villagers maintain ties to their home villages and, in effect, send remittances and provide money for the villagers back home. Sometimes they return to the villages to visit relatives or to engage in hunting, fishing, and harvesting activities.

There are, however, numerous reasons that merely moving the villagers to town is not a simple solution. First of all, as mentioned previously, subsistence communities must be small in order to live off of what the land has to offer. Second, the Inupiat hunting and fishing skills don't easily transfer to making a living in nonindigenous villages. So what would the thousands of emigrants do for a living? Further, a move from subsistence is a move away from indigenous culture.

Perhaps the Newfoundland experience has a bearing on the Native Alaskan situation. Since the 1950s, Canada's Newfoundland government has moved a large number of small fishing village inhabitants into towns. This was done to improve lifestyles, reduce poverty, and improve government services through consolidation. Between 1954 and 1965, the state-supported program moved 115 villages with a combined population of 7,500. Between 1965 and 1975, 148 more villages were relocated. The promised jobs and other opportunities never materialized, and the moves resulted in major disruption of economic, cultural, and social structures. After the collapse of the cod fishery in 1992, even more villages in Newfoundland were abandoned.

And finally, there is a darker side to the Arctic equation. There is a strong tendency toward alcoholism among the Inupiat. This is well recognized among the villagers, which is why many villages voted to prohibit the possession of alcohol. It is easy to see a possible humanitarian disaster over the horizon if the communities break up and the residents move en masse into nonindigenous villages.

The Louisiana Native American tribe on Isle de Jean Charles was awarded the Gulf Guardian Award from the EPA in recognition of its resettlement efforts. The purpose of the resettlement, other than the obvious one of needing to escape sea level rise, was described in the award document: *The goal of the resettlement is to maintain and strengthen the tribe's safety, collective identity, social stability, and contribution to the region.* The goals of the Biloxi-Chitimacha-Choctaw tribe are precisely those of the Inupiat and other Native Alaskan tribes.

The Denali Commission

Charged with the responsibility of looking after the Alaskan villages is the Denali Commission, which is similar to the Appalachian Regional Commission in its goals—to help disadvantaged communities. Established in 1998 by Senator Ted Stevens, the Denali Commission has distributed over a billion dollars, funding infrastructure of various types, including electricity, oil storage tanks, and health facilities. Clearly, the future moves of the villages will require help from the commission.

The chance of receiving federal help for the native villages is slim indeed in the current political climate. The Corps of Engineers could respond but only if funded by Congress. The immediate fate of the native villages may well require an act of Congress to broaden the role of FEMA or some combination of agencies. What's happening in the immediate future to the Alaskan villages will be happening in a wide swath of coastal America a few decades down the road.

Another major threat to the Inupiat Eskimo society is the government's strong interest in opening the Arctic to oil and gas exploration and eventual production. This very much threatens the lifestyle, the food sources (including the native wildlife), and the well-being of these Alaska villages. Experience elsewhere has shown that when this industry arrives, a new culture arrives with it, along with outsiders, roads, oil rigs, pipelines, and all the societal ills one associates with boomtowns.

Native Hawaiians also face a disproportionate risk from sea level rise. After Queen Liliuokalani abdicated in 1895, the federal government, in 1921, established, with the Hawaiian Homelands Act, a 200,000-acre area on Maui, land reserved for native Hawaiian homesteading. Unfortunately, parts of these homesteads are at particular risk from sea level rise. The good news is that, unlike the native Alaskan situation, a local government report suggests making an inventory of the situation and coming up with a solution for the future well-being of the native settlements. According to Mo Moler, a spokesman for native Hawaiians preparing the area for the homesteaders, "These people don't want to live in the city, but in the country where their ancestors used to live," he said. "This is about going back to the basics, culturally, physically and spiritually."

Is There Another Way?

Perhaps some creative thinking is needed concerning the methods of moving Inupiat houses away from the coast. The story of King Island in the Bering Sea, abandoned in the 1960s, may offer one possibility. The island was abandoned because of the loss of population due to the World War II draft, rockslides, and a devastating tuberculosis epidemic. Once home to 200 people, it is now a spectacular ghost town. Most of the population were moved to a neighborhood in Nome and were granted thousands of acres for a fishing/hunting camp at Cape Rodney, northwest of Nome. This small group of villagers and their descendants maintain a fishing camp there, and the members have wisely moved structures back from the coast because of erosion. Could something like this be the solution for thousands of erosion-threatened Alaska Natives? Could they be relocated and granted hunting and fishing lands to help preserve their way of life?

Maybe we can learn a bit from Chiloe Island in Southern Chile and the residents' way of moving small buildings (the size of most Alaskan Natives' houses) around. When a house is threatened by erosion or something else, the whole community pitches in to move it. On land, houses are moved rolling over rounded logs, pulled by oxen. Over water, the buildings are held afloat by "bladders" and pulled by fishing boats. It works for them, and it must be orders of magnitude cheaper than moving structures using modern heavy equipment.

It is unreasonable to expect that oxen be brought up to the Arctic Circle, but we are certain that, if the inhabitants approve, there could be much simplified and cheaper ways to go that the Corps of Engineers did not estimate, so that the total cost to taxpayers would be reduced. Of course, there remains the critical problem of building in a substrate of melting permafrost (unless they can find conveniently located rock outcrops upon which to anchor structures).

A final note indicating that the chance of receiving federal help for the native villages is slim indeed in the current political climate: according to Steve Benen, writing in the *Washington Post*, scientist Joel Clement, who worked for the Department of the Interior, disclosed how climate change affects Alaska Native communities, specifically mentioning Kivalina, Shishmaref, and Shaktoolik. For speaking out on climate change, Clement was removed from his position by then Secretary of the Interior Ryan Zinke. Clement declared that his removal was in retaliation for "disclosing the perilous impacts of climate change upon Alaska Native Communities and for working to help get them out of harm's way."

The Lesson from the Arctic

For Arctic villages, the impact of climate change became clear only recently—in the last decade or two. And now they are faced with impossible costs. Had they known of the reality of the coming impact of climate change several decades ago, they might have responded by planning and finding funding, locating suitable sites for relocation, moving a few buildings at a time, and requiring any newly constructed buildings to be portable and to be sited away from the shoreline. The fast-moving events in the Arctic have a lesson for the rest of America.

The rest of the American shoreline should look to the future and recognize that if we wait until the streets are filled with water and house prices crash, everyone will have to move at once. If we wait too long, all coastal states and many coastal communities will be desperately and simultaneously seeking funding to respond to the rising sea. In this context, building of new beachfront buildings, still happening on the U.S. coast, is madness.

LORD WILLING
AND THE CREEK DON'T RISE

· SUNNY DAY FLOODING ·

Nuisance flooding is on the rise in coastal towns. It's happening more and more, and it's getting worse and worse. It's the shape of things to come—the first visible sign of sea level rise along the American coastline. It's called *tidal flooding* or *nuisance flooding* or *sunny day flooding*. No matter what you call it, it is happening because the sea level is rising. The more the sea rises, the wider and deeper the nuisance flooding will become.

We all know, or can imagine, what to expect from a hurricane-related flood. The water may be swift-running and very damaging. In contrast, tidal flooding is quiet, almost creepy. Suddenly one day it's there, a two-inch-deep layer across the road you must use to go to work, to school, or to shop. Eventually the sites flooded periodically will be inundated permanently.

If you are unlucky, the nuisance tide may be a foot deep on your road. The worst flooding comes with the king tides. *King tide* is a term used to describe the highest of high tides. They have even been immortalized in song by British singer Billy Bragg in *King Tide and the Sunny Day Flood*.

FIGURE 3.1 This is a 2012 photo of a king tide in Annapolis, Maryland, where nuisance (chronic) flooding is among the worst in the nation. As sea level rises, king tides will flood more and more of this city. *Photo by Amy McGovern, Creative Commons.*

They are also called *perigean spring tides* because they occur at the new or full moon when the earth, the moon, and the sun are aligned to create the maximum gravitational pull on the earth's surface. When this spring tide occurs at the same time that the moon reaches its closest point to the earth during its monthly orbit, a super spring or king tide occurs. Usually it happens three or four times a year.

The flooding water may be fresh water if it's occurring far up an estuary or brackish if it's a little closer to the ocean. If the water comes from the ocean or through a close-by inlet, it may be pure salt water with oceanic salinity, the bane of the brakes and metal parts of cars, bicycles, and the like. Some people drive through the water without necessarily realizing it may be harmful to their cars. Tidal flooding is not storm-related, although onshore winds at most locations make the flooding worse. Instead, this is a sea-level-rise problem. Nuisance flooding is on the rise because the sea is on the rise. If onshore wind or heavy rain does coincide with tidal flooding, the impact is worse.

King tides and nuisance flooding are fully predictable and are usually

FIGURE 3.2 A 2015 king tide at Pier 14 in San Francisco, California. As sea level rises, king tides will produce greater and more frequent flooding. *Photo Source: Dave R., The California King Tides Project. Attribution noncommercial.*

mentioned in the local news outlets. You can particularly depend on local news outlets to keep you informed about coming king tides in Annapolis, Norfolk, and Miami.

Why worry about a few inches or even a foot of water on the street? Hamed Moftakhari and his coauthors argue in a 2017 research article that frequent tidal flooding has a significant effect cumulatively, a result of long-term repetitive occurrences. Among the cumulative effects is a reduction in the value of nearby property. And frequent tidal flooding can impact public surface transportation by causing delays and lost trips, affect road surface quality and sewer and drainage systems, and may be a threat to public health. Cleanup, road repair, and traffic control efforts by local governments are usually required.

Moftakhari and his coauthors note that nuisance flooding, taken as a whole, may be more damaging than occasional, infrequent storms. It is important to recognize this and not spend all the municipal effort and money on protection from the big one.

Joshua LeMonte and his five coauthors (2017) have observed that changes in the nature of groundwater caused by sea level rise and presumably by repetitive tidal flooding will impact the chemistry of soils, causing the movement of various undesirable elements. Of particular interest is arsenic, both naturally occurring and from human-caused pollution. With the intrusion of water from sea level rise, arsenic in soils is released into the water to be distributed to nearby soil surfaces and coastal waters. LeMonte et al. note that coastal soils containing arsenic are particularly prevalent in the Mid-Atlantic states.

The Urban Experience

In recent years, Washington, D.C., has experienced more than 94 hours a year of tidal flooding. Moftakhari and associates predict the city may have as many as 7,000 hours a year by the turn of the century (Moftakhari et al. 2017). At present, Baltimore and Annapolis, Maryland, have each shown a more than tenfold increase in flood days since 1960. Norfolk, Virginia, has the highest tidal flood level, averaging around 1.89 feet above the mean higher high-water mark.

With a degree of attention more like that given by a fishing village resident, Norfolk residents are well aware of high-tide stages and plan their routes, routines, and activities accordingly. In Norfolk, where the tidal flooding is particularly serious, the program from a Unitarian church routinely published tide tables and noted tidal flood stages to help people plan their route to church and which parking lot to use. The church recently sold its building and moved to higher ground.

In St. Petersburg, Florida, like any other coastal community, *no wake* signs abound on waterways to slow down boat traffic and thus reduce the size of waves striking the shore. But now, the same *no wake* signs are increasingly found on streets. This is done in a few communities in response to speeding vehicles pushing waves of tidal floodwater into adjacent yards and sometime into houses.

Perhaps the most famous and widely publicized example of sunny day flooding is that of Miami Beach, one of the nation's premier beaches. Those who view the flooding in person on Miami Beach often notice a very bad smell as sewers and storm drains are backed up and add to the flood, the odor, and the unseen disease-bearing pathogens. The problem

TABLE 3.1 Frequency of Nuisance Flooding: *Top 10 U.S. Communities with Greatest Average Increase*

	Average nuisance flood days, 1957–1963	Average nuisance flood days, 2007–2013	Height (feet)
Annapolis, MD	3.8	39.3	0 ft. 11.4 in.
Baltimore, MD	1.3	13.1	1 ft. 4.1 in.
Atlantic City, NJ	3.1	24.6	1 ft. 4.9 in.
Philadelphia, PA	1.6	12.0	1 ft. 7.3 in.
Sandy Hook, NJ	3.3	23.9	1 ft. 5.7 in.
Port Isabel, TX	2.1	13.9	1 ft. 1.4 in.
Charleston, SC	4.6	23.3	1 ft. 3.0 in.
Washington, DC	6.3	29.7	1 ft. 0.2 in.
San Francisco, CA	2.0	9.3	1 ft. 1.8 in.
Norfolk, VA	1.7	7.3	1 ft. 8.9 in.

SOURCE: National Oceanic and Atmospheric Administration (NOAA), 2014.

that Miami Beach has, however, is not much different than that of other seaside communities (see table 3.1).

With time, the level, the volume, and the frequency of nuisance flooding can be expected to increase, eventually reaching undesirable and undrivable water depths. Today, Miami Beach floods six times a year (it seems that news photographers must be on the scene for every flood). By 2045, based on a rapid sea-level-rise projection (six feet by 2100), Miami Beach can expect 380 floods per year, according to estimates by the Union of Concerned Scientists (UCS). That's more than one a day on average.

In recognition of this quickening hazard, Miami Beach has installed a $300 million floodwater pumping station, but the flip side of that is fear of polluting nearby Biscayne Bay, to which the water is pumped. Whatever the street has on its surface gets into the bay. Analyses of the water pumped into the bay with the new pumps indicated the presence of bacteria from both dog and human feces, the latter coming from leaky pipes and septic tanks. The longer-term fear is that eventually street floodwater will be pumped to the beach in large quantities, killing the tourist goose that laid the golden egg.

In addition, Miami Beach is building or raising some seawalls and raising the elevation of some of the streets that are typically flooded. This is costing perhaps another $200 million. The plan is to raise about half of Miami Beach's streets. This, of course, has an undesirable side effect, as such roads will act as dams during floods, pooling water into properties and neighborhoods that might otherwise have been spared. Already, the efforts have trapped some building entrances and equipment closets below sidewalk/street level, making them inaccessible or inconvenient and vulnerable to flooding.

Speaking of sparing buildings, many of South Beach's famous Art Deco buildings are at low elevation and certainly will be among the early victims of one too many floods. Serious consideration is being given to the possibility of raising the buildings, but the question under debate is whether a historic community is still a historic community when raised.

Oliver Milman, environmental reporter for the *Guardian*, notes that in great contrast to wealthy Miami Beach, Atlantic City, New Jersey, has little money available to address the flooding problem. Beyond the glitzy casinos and the much-beloved boardwalk is a community with a low tax base and many individual homes in flood-prone locations. Much of the effort to avoid floods is being made by individual homeowners raising their houses. Many people cannot afford that, however, as the cost of raising a single home here is in the range of $80,000 to $100,000. The city is considering raising all the streets and houses in some neighborhoods, but nothing of this magnitude is underway. There is also the consideration of putting up some new seawalls and preventing construction of new homes in readily flooded areas. Funding is simply not available.

To sum up the difference in the response to sea level rise between Miami Beach and Atlantic City, Miami Beach is doing a great deal, and the process consumes a huge chunk of their budget. Atlantic City is doing very little but is planning to do something as soon as funding may be found. It's not in the bank yet.

The Union of Concerned Scientists, in its detailed 2017 and 2018 reports of tidal flooding, defines *chronic flooding* as tidal flooding that occurs an average of 26 times per year (every other week). A community that is flooded at least every two weeks over 10 percent of its area is *chronically inundated*. In chronically inundated communities, retreat or substantial mitigation responses will certainly be required.

The following is a list of some of the Union of Concerned Scientists' expectations of U.S. tidal flooding, taken from its 2017 report:

- By 2035, nearly 170 communities—roughly twice as many as today—will reach the chronic flooding stage. Seventy percent of these will be in Louisiana and Maryland.
- By 2060, 270 U.S. coastal communities will have become chronically inundated.
- By 2100, 490 communities—including 40 percent of all East Coast and Gulf Coast communities that border the ocean—will be chronically inundated.
- Given a rapid rise in sea level, more than 50 important cities, including Oakland, California, Miami and St. Petersburg, Florida, and four of the five boroughs of New York City will suffer chronic inundation by 2100.

In the 2018 UCS report, again assuming a rapid rise of up to 6.5 feet by 2100, the states of Florida and New Jersey have the most to lose. "Within the next 30 years, roughly 64,000 homes in Florida and 62,000 in New Jersey will be at risk of chronic flooding. Along the Florida coast, Miami Beach alone, with its iconic high rises located within steps of the beach, accounts for more than 12,000 of these homes." In New Jersey, "Ocean City tops the list with more than 7,200 at-risk homes."

The 2018 report notes two housing risk hot spots. By 2045, Suffolk, Nassau, and Queens counties on Long Island, New York, could have 15,000 homes, valued at $7.7 billion and occupied by 40,000 people, at risk of chronic inundation. In the same time frame, the nine counties surrounding San Francisco Bay are expected to have 13,000 properties at risk valued at $8.6 billion and occupied by 33,000 people. Affected San Francisco Bay towns will likely include San Rafael, San Mateo, and San Jose.

A startling conclusion by the UCS (2018), again based on a rapid-rise scenario in the contiguous United States, is that *in the next 15 years* "roughly 147,000 existing homes and 7,000 commercial properties—currently worth $63 billion—are at risk of being inundated an average of 26 times per year, or more." About 280,000 people are estimated to live in these homes, many of which will need to either adapt to regular floods or, more likely, relocate.

It is important to know the assumed sea-level-rise rates behind sea-rise projections. In all the projections from the Union of Concerned Scientists, the assumption is made that the sea level rise will be rapid enough to cause a rise of 6.5 feet by 2100. This number is higher than the current scientific consensus of 3 feet by 2100. But the current acceleration of melting of the ice sheets, particularly in West Antarctica, means the higher estimate is a cautious but reasonable bet.

A busy season is ahead for nuisance flooding. The locations where flooding is occurring today are a sure sign of low-lying property that is vulnerable to future storm surges. The two-inch layer of water that covers your road today will be a one-foot layer in three or four decades, and perhaps three feet or more in eight or nine decades. For those who are skeptical about rising sea level, a visit to Norfolk, Virginia, during a king tide is in order. It will be a memorable experience.

DIRTY WATERS
AND WORRIED MINDS
· HEALTH CONCERNS IN AN AGE
OF CLIMATE CHANGE ·

Warming temperatures are most noticeable in the northernmost regions of our planet, and Siberia's Yamal Peninsula, which extends into the Arctic Ocean, is no exception. In 2016, temperatures reached as high as 95 degrees Fahrenheit on the peninsula, where summer temperatures usually hover around 60 degrees. In 2014, three massive sinkholes appeared—climate change phenomena attributed to exploding methane gas. In 2016, warming temperatures on the Yamal Peninsula melted reindeer carcasses and the corpses of the ancestors of the Nenets people, an ethnic group native to the Far North region of Russia. The melting permafrost exposed more than frozen reindeer corpses. It also exposed and unleashed anthrax that had been trapped in the ice for nearly a century. Scientists theorize the released anthrax seeped into groundwater and infected reindeer, killing over 2,000 of them before spreading into the human population. After the 2016 thawing event, anthrax took the life of a 12-year-old Nenets boy and sent at least 62 others to the hospital.

In a 2018 *Harper's Magazine* article, Moscow-based writer Noah Sneider notes that scientists have discovered fragments of the virus behind the 1918 Spanish flu pandemic in mass graves in the Alaskan tundra.

In addition, French scientists have found still-active viruses lurking in a 30,000-year-old chunk of permafrost. Sneider writes of an "uncanny symmetry" between the threats from thawing epidemics and the Nenets concept of understanding disease origins. "Nenets shamans speak of a world split into three realms: the upper world is home to Noom, the creator spirit; the middle world is the earth itself, home to humans and animals; the lower world belongs to Nga, Noom's brother, the spirit of death and disease. This lower world consists of seven layers of ice, and during times of trouble, the spirits of sickness rise from the frost."

A rising sea level and increased tidal flooding will aggravate current health challenges and create new ones. In 2014, the Intergovernmental Panel on Climate Change (IPCC), established by the United Nations and the World Meteorological Organization, listed with "high confidence" the "long-term risks to human health" from sea level rise and flooding:

- Death, injury, ill health, or disrupted livelihoods in low-lying coastal areas
- Severe ill health and disrupted livelihoods for large urban populations
- Extreme weather events (e.g., flooding intensified by sea level rise) leading to the breakdown of infrastructure networks and critical services such as electricity, water supply, and health and emergency services
- Food security and the breakdown of food systems

Who Is at Risk

Some populations will be more vulnerable than others. Typically, in the U.S., groups most at risk for health problems or death from climate change impacts include children, pregnant women, immigrants, non-English speakers, those with low income, those with disabilities or chronic health conditions, and the elderly. About half of the estimated 971 people who died in Louisiana (mostly in New Orleans) during Hurricane Katrina were over the age of 74. Drowning was responsible for 40 percent of the deaths, injury and trauma 25 percent, and heart conditions 11 percent.

The University of Miami Miller School of Medicine carried out a study in 2016 to determine sites and communities in southeast Florida that were at greatest risk for the "adverse health effects from the coming sea level rise." Perhaps surprisingly, at least in the case of South Florida, the wealthier populations were found to be most at risk. That's because in South Florida, large numbers of wealthy individuals live close to the shoreline where the rising sea and tidal flooding will have the most severe effects. Also, the wealthy in South Florida are often older retirees with preexisting health problems adding to their level of health risk.

Access to Emergency Help

Nuisance flooding, with increasing frequency, will flood and block low-elevation streets near the water, surrounding the residences of individuals and impeding access to emergency health care. As previously mentioned, the Texas Medical Center in Houston was dry but, due to flooded roads, was cut off from all access and egress during Hurricane Harvey. In this case, the cause was a major hurricane and not tidal flooding, but with time, as tidal flooding deepens with the rising seas, that hospital will remain a fitting example of a serious future problem.

Flooding also challenges emergency medical services personnel who could be unable to reach people in the community who need emergency care. The problem of lack of ready access to medical help will also occur in other lowland coastal areas such as the North Carolina Inner Banks, the Mississippi Delta, and South Florida. An example of this type of dilemma is the sole access road leading to Tybee Island, Georgia (year-round population 3,000-plus), which is readily flooded during highest tides on windless days.

Mental Health and Other Health Impacts

The adverse effects of repetitive tidal flooding that will occur with increasing frequency and magnitude may cause mental stress in vulnerable populations. Imagine the stress of repeated episodes of not being free to leave your house or neighborhood, missing work, school, movies, or church services because of tidal flooding, or the stress of watching your yard die from repeated exposure to salt water. This type of situation could lead people, mentally stressed or not, to decide to join the stream of refugees to higher ground.

According to a 2016 health summary publication by the U.S. Global Change Research Program, "Mental health consequences of climate change range from minimal stress and distress symptoms to clinical disorders such as anxiety, depression, post-traumatic stress [PTSD], and suicidality."

Many cases of depression and PTSD have been reported in Puerto Rico following Hurricane Maria (2017), undoubtedly exacerbated by wrecked homes and blocked roads, but also by the number of months during which much of the island lacked clean water, electricity, and communications. Puerto Rico was already struggling with an increase in mental illness amid a 10-year recession that brought soaring unemployment and increased poverty, as well as increased family separation caused by migration to the mainland U.S. Public health officials and caregivers say that Maria exacerbated these problems and will likely force many people to permanently leave their homes. U.S. Census data estimates 3.9 percent of the total population of Puerto Rico left the island in the year following Hurricane Maria.

The situation in Puerto Rico in 2017 and 2018 following the devastation of Hurricane Maria revealed lessons to be learned about numerous health problems that can occur because of a destructive storm and flood causing loss of access to hospitals, which are in darkness because of a long-term loss of power.

During emergencies, dialysis patients are given priority for treatment. In Puerto Rico, 5,600 patients depend on kidney dialysis three times a week. After Hurricane Maria, lack of electricity, transportation, fuel, and clean water all contributed to a health crisis for everyone in the region, not only for dialysis patients.

Hospitals without power were overwhelmed with patients seeking care. The true number of dead is not yet known. The original official number of dead was listed at 68. The *New York Times* estimated, by comparison with death numbers in "normal" years, that more than 1,050 people died during the hurricane and its aftermath. A study published in the *New England Journal of Medicine*, however, estimated the number of hurricane-related deaths in Puerto Rico was actually over 4,600. At best, the diverging numbers are a stunning example of postdisaster confusion, lack of accurate reporting, and a not-too-efficient government dealing with the situation. Even worse, these higher numbers reflect a government attempting to downplay or to cover up the shocking numbers of dead and an inept or inadequate federal storm response. On August 29, 2018, the Puerto Rican government declared the final death toll was 2,975 souls.

Moreover, conditions were perfect for waterborne and mosquito-borne diseases such as Zika, chikungunya, dengue, and even cholera, which is a bacterial illness easily spread when sanitation facilities are inadequate. The true level of such diseases poststorm on Puerto Rico is unrecorded, as public health lab testing was interrupted by the storm damage.

A very common observation (and complaint) after flooding is mosquitoes—lots of them. Although in the continental United States, flood-related mosquitoes rarely carry transmittable diseases, they are a serious nuisance and bring forth calls for widespread application of insecticides, which are themselves dangerous to the health and can pollute waters. One disease that is spread by mosquitoes and has been found in Florida but is particularly prevalent in Puerto Rico is Zika. In an October 2018 *Reveal* podcast segment called "The Storm after the Storm," reporter Beth Murphy unveiled that doctors on the island were outraged that the Puerto Rican government declared the Zika crisis to be over after ending the free testing of pregnant women, apparently to avoid further loss of tourism. In 2017 alone, before Hurricane Maria hit in late September, 1,500 women on the island had been diagnosed with Zika. Post-Maria, the Puerto Rican government reported that there were no new cases and declared the health crisis to be over, but it appears that they had merely stopped testing for the virus.

Snakes displaced by floodwaters are another common nuisance after a storm on the mainland U.S. After Hurricane Camille devastated the Mississippi coast in 1969, author Orrin Pilkey recalls seeing first responders carrying clubs to fend off snakes while inspecting houses, in search of survivors. Other wildlife may also take refuge in homes during storms. A Houston man returned to his home flooded by Hurricane Harvey to find a 10-foot alligator in his living room.

FIGURE 4.1 Floodwaters such as these in New Orleans following Hurricane Katrina (2005) are bound to be contaminated with various toxins and pathogens that cause a variety of gastrointestinal illnesses. In addition, emergency medical personnel will likely have difficulty reaching people in need. *Photo by Marty Bahamonde/FEMA.*

Less publicity has been given to the plight of the American Virgin Islands after the 2017 season's hurricanes. Category 5 Hurricane Irma hit all three islands, St. Croix, St. John, and St. Thomas. The islands' two hospitals were badly damaged and essentially nonfunctioning. More than 1,000 medical evacuees were sent off-island, most to Atlanta, but some to other mainland cities and to Puerto Rico. Those sent to Puerto Rico were forced to evacuate again by Hurricane Maria, which arrived two weeks after Irma. Many of the evacuees had lost their medical records, which had been attached to pouches around their necks, because they were moved under windy, rainy conditions. Upon arrival wherever they ended up, many had no identification, no money, no shoes, no relatives to look after and comfort them, and only the clothes on their backs. Governments at all levels in the Virgin Islands were medically unprepared for a big storm.

Pollution

Another major health hazard is caused by flooding damage to water purification plants, water wells, and sewage treatment plants. The various toxins and pathogens likely to be in any kind of floodwaters, including nuisance floods, can cause a variety of gastrointestinal illnesses resulting mostly in mild but sometimes severe flu-like symptoms. Many communities also have leaking sewer pipes and septic tanks that contribute to the pollution of floodwaters. As a tidal or nuisance flood recedes, contaminated soil remains behind to be further contaminated in the next tidal flood. Dysentery, diarrhea, hepatitis, and cholera are among the diseases borne in water mixed with sewage. Wounds are particularly susceptible to infections from contaminated water.

Furthermore, as the seas rise, increased tidal flooding can encourage the growth of mold due to excess dampness that can create or aggravate respiratory illnesses such as asthma. Intrusion of salt water into the groundwater drinking supply can increase the risk of hypertension and diarrheal disease. Warming seas may also increase Red Tides (algal blooms such as those that hit the West Coast of Florida in 2018), posing breathing difficulties for some people and threatening marine life.

It's clear that flooding, whether during sunny or stormy days, particularly with the rising seas, is apt to cause medical and psychological problems for the affected populations. Just as tidal flooding, storm surge flooding, and permanent inundation from sea level rise will increase in coming decades, so will the associated health problems.

THE FRONT LINE IN THE BATTLE

· THE U.S. MILITARY ·

By failing to prepare, you are preparing to fail.
BENJAMIN FRANKLIN

Among the governmental organizations, including the Congress and the Executive Branch, when it comes to concern about global climate change and sea level rise, the military may well lead the way.

In a *Forbes Magazine* article entitled "Does Our Military Know Something We Don't Know about Global Warming?," James Conca stated that the "environmental consequences of climate change will certainly cause future clashes and wars, and the countries and regions posing the greatest security threats to the United States are among those most susceptible to the adverse and destabilizing effects of climate change." Many of these points of instability, including North Korea, the Middle East, South Asia, and much of Africa, are already unstable and have little economic or social capital necessary for coping with the additional disruptions they will face due to problems related to climate change. Indeed, it is believed that some conflicts in Sudan and Syria are related to climate change–driven droughts.

Writing in the *Guardian*, Oliver Milman, Emily Holden, and David Algren argue that changes in weather patterns due to climate change are driving Central Americans to emigrate to the United States. They point out that a third of the jobs in the region are directly linked to agriculture, so droughts, heavier rains, or other changes in weather expected from climate change can have a major impact. While people may be immediately pushed northward by violence in the cities, in many cases these folks were initially displaced from the country due to crop failures. Conflicts from refugees will only get worse with climate change. Milman et al. wrote that the World Bank estimates that warming temperatures and extreme weather will force an estimated 3.9 million climate migrants to flee Central America over the next 30 years.

The Defense Department noted in a 2016 report sent to Congress that "global climate change is a multiplier that will aggravate problems such as poverty, social tensions, environmental degradation, ineffectual leadership, and weak political institutions." For example:

- When a drought destroyed much of the Russian wheat crop, wheat was no longer exported to the Middle East, as a result of which the price of wheat went up, which led to an economic crisis in Egypt.
- In Northern Mali, insurgents took advantage of increasing desertification and food shortages (2015) to supply food in exchange for Jihad agreements.
- A huge influx of refugees from Mali to Mauritania (2012) precipitated mass protests and violence in Mauritania.
- In the foreseeable future, Bangladesh will have 20 to 30 million sea-level-rise refugees from the Ganges-Brahmaputra Delta, with no apparent place to which to flee. Neighboring India and Myanmar are not interested in providing shelter and sustenance to the refugees.
- The already overcrowded Ganges-Brahmaputra Delta has received more than 600,000 Rohingyas, Sunni Muslim refugees forced out violently by the Buddhist-majority Myanmar military. This genocide threatens to be the foundation of a regional conflict.
- Higher temperatures and fewer winds bringing moisture from the Mediterranean caused the extreme and unprecedented drought in Syria between 2006 and 2009. Those conditions were likely a product of global climate change and one of the

major causes of the ongoing devastating war in Syria in which the U.S. military is involved. The Syrian drought forced large numbers of farmers, as many as 1.5 million, to move to the cities, creating poverty, social tensions, and little employment for youth vulnerable to radicalization and recruitment by ISIS. The Assad regime compounded problems by cutting some subsidies to farmers, fostering dissatisfaction and economic disruption.

It is likely that the U.S. military will continue to be involved in more regions of instability in the future in conflicts that derive from climate-change-related crises. Initially, these conflicts may be drought-driven, but eventually conflicts may arise from emigrants who move in large numbers away from the waters of the rising seas.

Politics in the Mix

The U.S. Congress does not always see eye to eye with the military. A particular flashpoint for the Republicans in Congress is the twelve-page 2016 Department of Defense Directive (DoD 4715.21). It is a dense, jargon-filled document that outlines the tasks and responsibilities of various undersecretaries, assistant secretaries, and component heads for advancing "climate change adaptation and resilience" in the military. The basic problem in Congress is disbelief or outright denial, mostly on the part of Republicans (and the president), about the importance of global climate change.

In 2016, Republican Congressman Ken Buck of Colorado introduced a particularly absurd amendment that blocked funding of any consideration of global climate change by the military. Congressman Buck observed "President (Obama) has talked about an increase in the climate temperature on the planet. It is a fraction of a degree every year. How that is a current threat to us is beyond me. The military, the intelligence community, and the domestic national security agencies should be focused on ISIS and not climate change. The fact that the President wants to push a radical green energy agenda should not diminish your ability to counter terrorism." Ironically, ISIS can trace its origins in part to the effect of climate change on farmers, as discussed earlier.

While a number of conservative pundits take the view that the Pentagon is being shackled or weakened with climate change directives, the military itself, with its global reach, thirst for weapons systems, and seemingly unending wars, is also clearly part of the climate change problem. Desiree Hellegers writes in *Counter Punch* magazine that the massive U.S. military may be the "largest single driver of climate change worldwide." She argues that since the military may be the single largest oil consumer (359,000 barrels per day in 2009), it is "the largest polluter in the world" and that by contributing greatly to greenhouse gasses and hence climate change in general, "the U.S. military poses a significant threat to the U.S. military."

Sea Level Rise Impacts at Military Bases

Perhaps the best example of an immediate climate change challenge for the military is the need to raise the U.S. Navy's docks and piers to accommodate the rising sea. Naval Station Norfolk in Virginia, the largest naval base in the world, was established 100 years ago when sea level was 1.5 feet lower than at present. The area is very flat, much of it on landfill that is currently compacting (and subsiding), which adds to the relative sea level rise. The sea also is rising relatively rapidly here because the area is located on a sinking glacial upward fore-bulge that was created during the last ice advance by the immense weight of the glaciers miles to the north. Ice age glaciers placed so much weight on the land it occupied that it caused lands in front of the glaciers to bulge up. With the weight of the glacial ice long gone, this glacial fore-bulge is now sinking at a rate of 1.5 inches per decade. The compressing fill and the lowering glacial fore-bulge, along with the piling up of waters due to a slowing Gulf Stream (at its weakest in 1,400 years), all combine to give the Norfolk region one of the highest rates of sea level rise in North America.

The reason a slowing Gulf Stream raises sea level along the Atlantic shore is that much of the Gulf Stream at depth flows to the right toward the center of the ocean. This raises a mound of water in the center of the ocean basin. When the Gulf Stream slows, the force of water flowing to the ocean center slows, and the mound begins to subside. This pushes

FIGURE 5.1 One of the sea level hot spots on the East Coast of the United States is at Norfolk, Virginia, where the largest naval base in the world is located. The high rate of sea level rise here is partly due to the fact that the land is sinking. To accommodate the sea level rise, the U.S. Navy will have to raise the docks and also help protect the city of Norfolk, where many of the base employees live. In this photo, sailors man the rails as the amphibious transport dock ship USS San Antonio (LPD 17) leaves port. *Photo credit: U.S. Navy photo by Mass Communication Specialist 2nd Class Adam Austin.*

water toward the continent, causing an increase in sea level in a landward direction.

There is general agreement that an additional 1.5-foot sea level rise will be a tipping point beyond which critical damage will occur to the naval base. In addition, the military must be concerned with protecting Norfolk and other nearby towns where workers and many sailors live.

Dr. Joe Bouchard, former Naval Station Norfolk base commander, notes that the base needs a complete overhaul "to avoid catastrophe." The situation, according to Bouchard and others, is one of lots of planning but little real action. Over the past fifteen years, four piers were raised, making them double-deckers, but this was done to replace aging structures and to raise various utility lines out of reach of nuisance floods. So, as of 2018, the world's largest naval base has not responded significantly to sea level rise, a clear threat to future combat readiness.

According to a recent report by the Union of Concerned Scientists, the naval installations likely to face daily "nuisance flooding" in coming decades are:

- Maine, Portsmouth Naval Shipyard
- New Jersey, Coast Guard Station, Sandy Hook
- Maryland, the U.S. Naval Academy, Annapolis
- Washington, D.C., Naval Support Facility Anacostia
- Washington, D.C., Washington Navy Yard
- Virginia, Naval Station, Norfolk
- Virginia, Naval Air Station Oceana/Dam Neck Annex
- Georgia, Naval Submarine Base, Kings Bay
- Florida, Naval Station, Mayport
- Florida, Naval Air Station, Key West

But the potential national security ramifications of global climate change extend far beyond the need to raise the docks at naval bases and all the rest of the world's ports used by the U.S. Navy. Bases of other military branches at similar risks to daily nuisance flooding (and eventually inundation by sea level rise) include:

- Coast Guard Station, Sandy Hook, New Jersey
- Joint Base Anacostia-Bolling, Washington, D.C.
- Joint Base Langley-Eustis, Virginia
- Marine Corps Base, Camp Lejeune, North Carolina
- Marine Corps Recruit Depot, Parris Island, South Carolina
- Marine Corps Air Station, Beaufort, South Carolina
- Joint Base Charleston, South Carolina
- Eglin Air Force Base, Florida

The importance of these threatened facilities is illustrated by the summary of the various functions of Joint Base Charleston (JBC): "Joint Base Charleston is one of 12 Department of Defense (DoD) Joint Bases and is host to over 60 DoD and Federal agencies. The 628th Air Base at the Charleston facility delivers installation support to a total force of over 90,000 Airmen, Sailors, Soldiers, Marines, Coast Guardsmen, civilians, dependents, and retirees across four installations, and has 22 miles of coastline totaling almost 24,000 acres."

MacDill Air Force Base in Tampa, Florida, is at low elevation and is the headquarters of the U.S. Central Command from which the wars in Iraq, Syria, and Afghanistan are controlled. This important base, susceptible to the rising seas, is also the headquarters of the U.S. Special Operation Command.

In October 2018, Hurricane Michael's 155-mile-per-hour winds tore through Tyndall Air Force Base outside of Panama City, Florida. During a Senate Armed Services Committee hearing in December 2018, Senator Tim Kaine placed the price tag for damage at Tyndall Air Force Base at around $5 billion dollars. Hurricane Michael's unexpected ferocity is a harbinger of the increased intensity of storms expected from our changing climate.

Other conclusions of the Union of Concerned Scientists' 2016 report are:

- By 2050, most of these installations will see 10 times the number of today's floods.
- By 2070, half of these sites will receive more than one flood daily.
- By 2100, eight bases are likely to lose 25 to 50 percent of their land area.
- Many surrounding communities will also be subjected to flooding, limiting access to military facilities.

Not mentioned in the report is the extreme sea level threat to the Marshall Island Pacific atolls, which host the Ronald Reagan Ballistic Missile Test Site. All told, the U.S. Navy has facilities of various kinds in 20 countries, including Japan (6), Korea (1), Bahrain (3), Diego Garcia (1), Italy (6), United Arab Emirates (2), Saudi Arabia (2), and 13 bases in other countries.

Why the Military's Concern?

Former deputy undersecretary of defense for environmental issues Sherri Goodman states that the military's great concern about climate "shouldn't be thought of as spending money on climate change. It's an investment to protect our troops." She said she wouldn't want her son or daughter to

FIGURE 5.2 The commercial Port of Long Beach, California, like other U.S. ports, has a large area at low elevation adjacent to the docks for storage of containers and other shipped materials. Raising docks and protecting freight storage areas will be a major undertaking for commercial entities and for the Navy, which has port facilities all over the globe. *Photo source: Google Earth.*

be unprotected when our military bases are affected by sea level rise and coastal subsidence.

Sea level rise, because of its huge impact on the world's ports and docking facilities, is an immediate concern—so immediate that the military must confront and sometimes ignore the climate change skeptics in order to fulfill their mission of maintaining military readiness.

A huge concern is the cost of maintaining, repairing, moving, raising, replacing, or abandoning naval facilities around the world. The cost will be one of the major immediate burdens to the U.S. Treasury when the nation finally faces up to its climate future. Recognizing the costs will very likely result in some major shakeups, changes, and perhaps shrinking of America's global security obligations.

It is the obligation of our military to not be caught unawares and to be prepared for all imaginable, discernable eventualities in this rapidly changing and evolving world. Clearly the actions of the American military establishment demonstrate an impressive understanding of the significance of global climate change in terms of potential future military challenges.

AT-RISK COASTAL ENVIRONMENTS

· IS RESILIENCE FUTILE? ·

The impact of sea level rise on the earth varies considerably according to the affected geologic environment and the topography. Unquestionably, gentle coastal slopes are more at risk to inundation than steep slopes.

On a continental scale, the East Coast (trailing edge) of North America is riding along on the westward-moving North American tectonic plate. New plate is forming all the time at the Mid-Ocean Ridge out in the Atlantic. This results in a *passive continental margin*, where everything is moving along peacefully. The region consists mostly of gently sloping coastal plains, lined by loose, unconsolidated sand derived ultimately from the Appalachian Mountains. Such passive margins are particularly vulnerable to sea level rise.

On the mountainous West Coast of America, the leading edge of the North American plate is no longer moving peacefully along. Instead, it is clashing violently with the Pacific plate, creating earthquakes and pushing up the strata to form mountains (the Rockies, the Cascades, and others) along the edge of the continent. This is an *active continental margin*. Such shorelines are steeper, often rocky, and will be less affected or at least more slowly affected by sea level rise.

FIGURE 6.1 Cedar Island, Virginia, a barrier island on the flat coastal plain along the Delmarva Peninsula, is one of 14 islands owned by the Nature Conservancy. In the 1950s, developer Richard Hall platted 2,000 lots to constitute a future "Ocean City." Erosion halted the plans. Today there are no houses remaining on Cedar Island. The remains of the stilts from one long-gone house are on the right. *Photo courtesy of David Harp © 1998/ChesapeakePhotos.com.*

Another difference between mountainous and coastal plain coasts, or active and passive coasts, is the effect of dams on the all-important supply of sand carried by rivers to the beaches at the shoreline. On active coasts like the U.S. Pacific Coast, most rivers dump their sediment load directly into the ocean and sometimes directly onto the beaches lining the shore. Shoreline retreat (erosion) is exacerbated when dams on rivers trap sand behind them, thus halting the flow of sand to the beaches. In contrast, on coastal plain coasts, the river sand load is normally deposited at the heads of estuaries, miles from the beach, so the dams on rivers have much less, if any, impact on beach sand supply—with the notable exception of the Mississippi River.

Slope of the land is also a critical determinant of erosion potential in a rising sea. Obviously, a shoreline hit by a rising sea will retreat faster across a gently sloping plain than across steep hills and mountains. In contrast to the gently sloping Atlantic and Gulf coasts, the Pacific Coast

FIGURE 6.2 On a mountainous coast such as seen in this California example, sea level rise will have a much smaller or at least a much slower effect on coastal communities. In contrast, for every one foot of sea level rise on coastal plain coasts of the Atlantic and Gulf, the shoreline, in theory, may move back thousands of feet. *Photo courtesy of Miles Hayes.*

of the United States is much more variable in its nature. Although there are many short reaches of the Pacific Coast that are relatively low lying, it is generally much steeper than the Atlantic and Gulf coasts.

Both types of coasts have extensive high-risk development lining the shores, along with the requisite infrastructure (roads, water, sewer, toxic waste sites, and power plants, some of them nuclear) that is at risk from inundation by the rising seas.

Atolls

Overall, about a million people call atolls home. These sometimes-beautiful features, with palm trees and warm breezes, are found in both the Indian and Pacific oceans. The rising sea immediately threatens them, and because of this they can be accurately characterized as the "canaries in the mine." They may well be the first natural environments to be entirely abandoned by humans. In most cases, the actual living area on atolls is three to six feet high, an elevation easily invaded by storm waves. Most atolls have a ridge of storm-tossed coral reef fragments facing the open ocean that acts as a protective natural beach or "seawall" against the effects of minor storms.

The United States possesses only eight atolls, all in the Pacific. These are Baker Island, Howland Island, Jarvis Island, Johnston Atoll, Kingman Reef, Midway Atoll, Palmyra Atoll, and Wake Island. Most are uninhabited or lightly inhabited, and Palmyra Atoll is the only incorporated territory. None of these have any permanent population, just temporarily stationed military or scientific personnel. Hence there will be no American atoll refugees, and the fate of U.S. atolls is of little national consequence, except for the military. It is expected, however, that our close relationship with the 50,000 people of the multi-island Marshall Islands will someday result in them joining the flight of sea-level-rise refugees to America.

Arctic Shorelines

The Arctic shoreline (North Slope of Alaska and the Chukchi Sea) problem mainly affects small subsistence villages (populations of 400 to 700) of Native Alaskans. The plight of these hardy people is discussed in detail in the Inupiat refugee chapter (chapter 2). The total number of people immediately affected here is perhaps fewer than one hundred thousand. Sea level rise combined with melting permafrost in beaches and lessening protection of beaches by seasonal sea ice will eventually drive all these small villages inland. The villages must be moved.

Moving villages to a site close to traditional fishing or hunting grounds is a costly process because of the problem of construction in melting permafrost zones. As in the case of the world's atolls, we expect all the Arctic beachfront villages will be moved or destroyed within the next 50 years.

Rocky Shores

This large category encompasses a wide variety of shoreline types typified in the United States by the coasts of Maine, New Hampshire, Connecticut, Rhode Island, California, Oregon, Washington, and Alaska.

FIGURE 6.3 An eroding bluff in Isla Vista, California, with a long row of seriously threatened houses. The beach is exposed bedrock at low tide. A USGS study by Limber noted that southern California cliffs will recede considerably during the rest of this century as the sea keeps rising. At the top part of this cliff is a concrete apron, a futile attempt to keep erosion under control and the houses from falling in. Further down the beach, one can see a seawall. *Photo by Alex Snyder, U.S. Geological Survey.*

The beaches range from boulder-strewn glacier deposits to short, sandy pocket beaches with rocky headlands at both ends of the beach to bluffs and cliffs of varying hardness and durability and susceptibility to erosion. The nature and susceptibility to sea level rise may change very quickly over very short distances, in contrast to the long beaches on barrier island chains. Rocky coasts can have spits, sandy shoreline reaches, and soft bluffs that sometimes erode as quickly as coastal plain shores. Erosion rates vary widely but are usually relatively slow along rocky coasts, and the impact of sea level rise will be much slower than on deltas and barrier islands. However, in a U.S. Geological Survey (USGS) study report, coastal geomorphologist Pat Limber noted that in Southern California, cliffs could recede more than 130 feet by the year 2100 as the sea keeps rising.

In heavily developed California, much construction atop bluffs is threatened by erosion, which will increase as the sea level rises. Often, erosion of bluffs is a principal source of beach sand, and "protection" of buildings perched atop the bluff, with sea walls at the base of the bluff, leads to erosion of the beach because of the lost sand supply. Seawalls are a common occurrence along California's coast, although the state is trying to limit them because of their detrimental effect on beaches. The combination of forest fires near coastal communities, followed by heavy rains, both phenomena likely to be increased by climate change, may lead to more catastrophic mudslides like those that affected the Santa Barbara, California, area in 2017.

The U.S. West Coast is a mixture of rocky shorelines separated and often fronted by beaches and spits. The population centers with significant numbers of buildings that will be threatened by sea level rise include San Francisco (San Mateo, Alameda, Santa Clara, and Marin counties) and Los Angeles (Los Angeles and Orange counties).

Deltas

Deltas are complex geologic features formed from river sediment contributed to a standing or slow-moving body of water. Deltas at the mouths of today's rivers were formed when the sea level rise slowed down about 5,000 to 6,000 years ago. They are shaped by waves and tides and influenced by the profile of the offshore seafloor, vegetation, river processes, and the volume and type of river sediment. Deltas are particularly desirable sites for development because of their rich agricultural land, easy access to the sea for fishing and aquaculture, and economy-building port facilities. All of these characteristics apply to the Mississippi Delta, the only major U.S. delta.

The Mississippi Delta and deltas all over the world have been increasingly altered by human activities. In fact, deltas are responsible for the highest immediate risk from sea level rise to the largest number of inhabitants of any of the earth's major geologic environments. The rising sea level, occurring simultaneously with an expected increase in intensity of storms, arrives on the scene at a time when many deltas are already in a state of degradation. This is in large part due to the high population density. Large cities built on deltas, including Shanghai, New Orleans, and Bangkok, are at particular risk of inundation.

The effects of sea level rise on deltas include increased flooding of cities, towns, and villages, loss of storm-surge-dampening wetlands, salinization of farmlands and groundwater, and rapid coastal retreat.

Delta size is a function of a battle between waves nibbling at the edges of the delta and the volume and grain size of the river's sediment load. For instance, most of the Amazon River Delta is below sea level because the river sediment load is primarily mud, which is readily moved offshore by waves. On the other hand, the River Nile contributes (or used to contribute before the Aswan High Dam was constructed) a larger-grained, sandier sediment load to a shoreline in the Mediterranean with relatively mild waves. So the delta extends well out to sea and accommodates both valuable farmland and large cities. The Mississippi has a sandy mud load that allowed, at least in the old days, the formation of large areas of marsh, sand bars, and islands that could be developed. The trapping of sediment behind large dams on the Missouri River, far upstream, has had a significant negative effect on the sediment supply to the Mississippi Delta.

The Mississippi Delta is three million acres (4,600 square miles) in areal extent, 3,800 square miles of which is estuarine marsh. It is the seventh-largest delta in the world. Many miles of human-made levees have mitigated floods for communities on the delta but have prevented the river from contributing fresh sediment and nutrients to the marshes during floods. Instead, today, much of the river's flow and sediment are directed across the delta and straight into the Gulf of Mexico.

Many miles of canals crisscrossing the delta, used for navigation, pipelines, and access to oil production facilities, have led to extensive and continuing erosion of marshes. Extraction of groundwater and particularly of oil and gas has contributed significantly to the subsidence of the Mississippi Delta's surface, which was already occurring as a result of natural compaction of the muddy deltaic sediment. Subsidence adds significantly to the relative sea-level-rise rate. On some portions of the Mississippi Delta, the sea level is rising at a rate of 4 feet per century, compared to a typical current rate of 1 to 1.5 feet per century globally.

Between the loss of sediment from upstream dams, the natural and induced subsidence, and the rising sea level, most of the Mississippi Delta will "drown" by 2100. Almost all of the roughly two million people who live on the delta will become climate refugees.

Barrier Islands

These are long, narrow sandy or gravel islands bounded by inlets on both ends, with the open ocean on one side and a lagoon on the other. They are 15 to 30 feet thick, ranging from less than 1 mile to more than 100 miles long (Padre Island, Texas, the longest American barrier island, is 135 miles long), and several yards to a few miles wide. Barrier islands are among the world's most dynamic natural environments, capable of migrating landward in response to sea level rise. Many formed thousands of years ago when sea level was lower (where the ocean is now) and migrated miles landward to their present locations as the seas rose.

Eighteen thousand years ago, sea level was 400 feet lower than it is now. By 5,000 years ago, after the sea had risen to within 1 or 2 feet of its present level due to the melting of ice sheets, island migration slowed way down. Many islands began to widen and grow upward as dunes formed from sand pushed ashore by the wind and waves. Today the process has reversed itself, and now many islands are thinning by erosion on both the ocean and lagoon shorelines, facilitating future island migration. In other words, putting it in human terms, the islands have sensed sea level rise and are getting ready to commence migration. And they do so by narrowing so that sand can readily be moved across them in storms.

Barrier island landward migration (and hence the survival of the islands in a rising sea) depends on storms. Migration occurs when storms cause the ocean side of the island to retreat (erode). At the same time, sand is picked up by the breaking storm waves and moved from the shore onto the island. In the case of large storms, the sand will move all the way across the island and into the lagoon. The net result of combined erosion on the front side and widening of the back side of the island by storm overwash is migration of the entire island in a landward direction. For this to happen, the islands have to be narrow enough for storm overwash to cross the entire island. The whole process is something like the movement of a tank tread.

There are 2,200 barrier islands around the world, found on every con-

FIGURE 6.4 Padre Island, Texas (the longest barrier island in the United States), along the Gulf of Mexico, illustrates why sea level rise will generally push shorelines back much farther on flat Atlantic and Gulf coastal plains than on steeper Pacific coasts (see figure 6.2). *Photo courtesy of Miles Hayes.*

tinent except Antarctica. The number of barrier islands will vary each year as new ones form when storms cut new inlets or when inlets close as waves move sand into them. Almost all barrier islands are situated in front of broad, low-lying coastal plains.

According to a survey by Matt Stutz, the number of islands in various countries ranges from 405 in the United States to 1 each in New Zealand, Cameroon, and Liberia. (See table 6.1 for countries with the largest number of barrier islands after the United States.)

Nothing beats a vacation in a beachfront cottage on a lovely barrier island. In the temperate zone, barrier islands are particularly desirable for development due to the easy access to the sea, the wonders of beach relaxation complete with shell hunting and fishing, and the much-treasured sea view. Shells have their own story to tell. The presence of oyster shells (and other shells that would normally thrive in lagoons behind the island) and marsh peats on the beaches of barrier islands is proof of island migration, provided the beach has not been nourished by pumping or trucking in new sand.

TABLE 6.1 Barrier Islands: *Countries with Largest Number*

U.S.A.	405
Russia (Siberia)	226
Australia	208
Canada	154
Madagascar	119
Mozambique	115
Mexico	104

SOURCE: Stutz and Pilkey, 2001.

Development on barrier islands can be extremely dense, often with long lines of high-rises jammed up against the beach, as seen in Miami Beach and Ft. Lauderdale, Florida, Virginia Beach, Virginia, Myrtle Beach, South Carolina, and Gulf Shores, Alabama, among dozens of other communities. In less-developed countries, such as those in tropical and arctic climate conditions, seafood extraction for local consumption, not tourism, is often the main island activity.

Ironically, sea level rise and storms, the main processes that make barrier islands hazardous for development, are the same processes that barrier islands need for survival. Holding a barrier island still in order to protect buildings will result, in a multi-decade time frame, in the loss of the island as the sea rises. In other words, islands held still by development cannot long survive without extensive, even heroic, engineering efforts. So who is going to win: Nature or the people who live on the islands? Let's just say Nature bats last at the shoreline.

The future of developed barrier islands in a rising sea is a grim one. We believe that a three-foot sea level rise will make it impossible to maintain a beach with beach nourishment, and high-rise-lined shorelines on islands will be held in place by massive seawalls on both sides of islands, with no beaches in front of the walls. A six-foot sea level rise will force total abandonment of many barrier islands. All told, we believe that the cost of many resilience efforts would be better spent on retreat. Retreat offers the added value of allowing the islands to survive and to migrate in response to the sea level rise.

Almost all East Coast and Gulf Coast barrier islands of the United

States are lined with buildings, with the important exceptions of national seashores and state and local parks. The Nature Conservancy, a nonprofit environmental group, has accomplished an unusual type of barrier island preservation. While pretending to be developers, they purchased fourteen undeveloped barrier islands on the Delmarva Peninsula and are now letting natural processes prevail on these islands.

Recently the U.S. Fish and Wildlife Service made an exciting pioneering decision to save a wildlife refuge for future generations. In cooperation with the Open Spaces Institute, the agency has accumulated four pieces of property on the mainland near Cape Romain, South Carolina. The largest one is 5,619 acres behind Bull Island, which is part of the Cape Romain National Wildlife Refuge immediately adjacent to the Francis Marion National Forest. The purpose of the purchase of this land is to maintain the wildlife refuge as the shoreline advances with sea level rise and as Bull Island, a barrier island, migrates over and away from parts of the refuge.

This is a brilliant example of working with Nature, an example for others who face a rising sea to follow. For example, a town could buy a large property on the mainland as insurance for the time when they must move back from the ocean. Moving buildings back could be accomplished on an individual homeowner basis or by moving threatened front-row houses back one after another.

Beaches

Beaches are extremely rich biologically, from the meiofauna, an assemblage of organisms that live in the microscopic spaces between sand grains, up to the birds and surf-zone fish that feed on the snails and clams that feed on the meiofauna. Left to its own resources, the beach ecosystem is essentially indestructible. Sea level rise will simply move the whole system to higher ground.

We are the principal enemy of the beach. We drive on it, spill oil on it, pollute it with our wastes from nearby farms and cities, and mine it for sand. Perhaps the worst thing we do to the beach is our reaction to the rising sea. We try to hold the dynamic, ever-changing beach in place by "stabilizing" it in order to save the buildings, the streets, the rest of the

TABLE 6.2 Surfrider's Beach Report Card

East Coast		Gulf Coast		West Coast	
Maine	B	Florida	D	Hawaii	C+
New Hampshire	B	Alabama	D	California	A
Massachusetts	B	Mississippi	D	Oregon	B-
Rhode Island	B	Louisiana	D	Washington	B
Connecticut	C	Texas	D	Alaska	D
New York	C+				
New Jersey	F				
Delaware	C				
Maryland	B				
Virginia	C				
North Carolina	D				
South Carolina	C				
Georgia	F				
Florida	D				
Puerto Rico	D				

The 2018 State of the Beach Report Card grades given to each U.S. coastal state by the Surfrider Foundation. Grades are based on the quality of beach management, especially the states' efforts in preservation of beach quality.

infrastructure, and all of the tourist paraphernalia. We stop it from moving by stabilization, either soft or hard or both. Beach nourishment, the introduction of new sand onto the beach to replace what the waves have taken away, is considered *soft stabilization*. Seawalls, rock revetments, bulkheads, groins, breakwaters, jetties, and other coastal engineering structures are considered *hard stabilization*.

In recognition of the fragility of beaches, the Surfrider Foundation issues annual report cards, grading the quality of beach management in each state. The Surfrider Foundation graded 30 coastal states, including Hawaii and Puerto Rico, on (1) their policies for protecting the beaches from coastal erosion, (2) irresponsible beach fill, (3) sea level rise, and (4) poorly planned coastal developments. The 2018 version grades ranged from A for California to F for New Jersey and Georgia (table 6.2).

Beach nourishment is the wave of the immediate future on all U.S. coasts. It currently is the most widely used technique to hold the shoreline still, but its prominence in coastal management will gradually fade with a rising sea. Lifespans of nourished beaches will gradually diminish as the shoreline position becomes increasingly unstable and out of equilibrium with the level of the rising sea. An unstable nourished beach is one with a short lifespan.

Nourishment is not without physical hazards. Nourished beaches erode much faster than natural beaches, and this rapid rate of shoreline retreat often results in a vertical sand scarp at the high-tide line. This scarp can be a cliff as high as 20 feet or more, becoming a hazard to beach walkers, especially at night. The most famous beach scarp victim was New Jersey Governor Jim McGreevey, who fell off a 4-foot-high beach scarp and broke his leg. Another less-common hazard is the formation of sand bars offshore with gaps that create rip currents where none had been seen before. Deaths caused by such currents have occurred on a number of nourished beaches, including Ocean City, Maryland, John Lloyd Beach, Florida, and Naples, Florida.

In addition, as more and more communities nourish their beaches, the demand on the federal budget that usually pays most of the cost will become intolerable to the taxpayer, and the cost will revert to individual towns and states. This transfer of monetary responsibility will markedly diminish beach nourishment around the nation. The taxpayer attitude is increasingly becoming, "I wasn't dumb enough to build a house next to an eroding beach, so why should I pay for a new beach to save the houses of those imprudent people?" But for the coming decade or two, nourishment will be the answer.

Most U.S. beach nourishment projects cost $300 to $1,000 per linear foot. One two-mile project on the Outer Banks (north of Rodanthe, North Carolina) cost nearly $2,000 per linear foot. Many beaches have been nourished multiple times. The most repeatedly nourished beaches in North Carolina have been nourished 5 to 10 times, but Pea Island is witness to 16 nourishment projects (for protection of Highway 12 only). East Beach, Santa Barbara, California, has been nourished 23 times; Fort Pierce, Florida, 17 times; Panama City, Florida, 10 times; and Harrison County, Mississippi, 8 times.

The actual lifespan of beaches is variable and of course depends partly on unpredictable storm schedules. Overall, the wave energy (wave height

TABLE 6.3 Beach Nourishment Histories

State	# Episodes	Cubic yards	Cost in 2018 dollars
New Jersey	325	186 million	$1.9 billion
North Carolina	254	123 million	$844 million
Florida	495	273 million	$2.4 billion
California	343	351 million	$281 million
Washington State	11	1.2 million	$16 million

Examples of the beach nourishment effort in several states. Clearly beach nourishment will become a major item in state and federal budgets in coming years.

SOURCE: *Beach Nourishment Viewer*, 2018. Program for the Study of Developed Shorelines, Western Carolina University.

on average) of a location is the main factor. In Florida, beaches often last 7 to 9 years. In North Carolina and New Jersey, where the typical waves are higher and storms more frequent, 3 to 4 years is a common lifespan. The 1981 nourished beach in Miami Beach lasted more than 15 years, possibly because of some natural cementation of the shelly (calcareous) sand in addition to relatively small waves because of the sheltering effect of the offshore Bahama Banks.

Most beach nourishment projects use sand dredged up from the seafloor. (Bringing in sand by dump truck is usually a much more costly approach.) The environmental impact of mining the continental shelf for sand and smothering the beach with new sand is discussed in chapter 7.

Passengers sitting next to the window on flights across the shoreline of Miami Beach to the Miami Airport can see a long, white band of discolored water, hundreds of feet wide. This happens because the original material from the 1981 beach nourishment was made up of fragments of shells, corals, and calcareous algae, all of which are soft and easily and continuously ground up by the breaking waves. This creates a serious problem for filter-feeding organisms and has been shown to eliminate them altogether in some cases.

The Bahamas Problem

The problem of surf-zone murky water caused by lots of suspended sediment in nearshore waters and beyond may increase dramatically in the future on Florida's beaches. The world in general and Florida in particular face a severe sand shortage. A hundred miles offshore from Florida's east coast lies the Bahamas, a nation with lots of sand that beckons to the hungry beaches of Florida. Fortunately, a 1986 U.S. law prohibits spending federal money to mine or transport the beach sand from other countries.

Needless to say, efforts are underway to change the laws, and if that happens the soft carbonate beach sand will likely be spread far and wide along Florida's east coast. Consequently, the problem of excess suspended sediment in the nearshore will also be spread far and wide.

Mining Bahamian sand will inevitably have unexpected side effects. For example, use of such sand would introduce nonnative species to Florida. And what will long- and short-term extraction of millions of cubic yards of sand do to the Bahamas? The depressions left behind will certainly affect wave height and wave refraction conditions. If the mining site on the Bahama Banks is not chosen carefully, it could prove very damaging to some Bahamian communities, leading to flooding and wave damage from storms. The Bahamian natives will be best served if they leave their sand right where it is and let the United States solve its beach nourishment problems some other way.

Hard Stabilization

Holding the shoreline in place is a costly process, particularly in a time of rising sea level. Basically, under most circumstances in a time of rising seas, the hold-the-line solution will not succeed in the long term and is extremely damaging to the nearshore ecosystem.

Perhaps we are at the beginning of what will probably become a future wave of beach management. Daniel Sweeney writes in Florida's *Sun Sentinel* paper that "instead of dumping ton after ton of sand only to see it wash away again, should we start making plans to cede our beaches to the ocean and build sea walls to defend what's left? Is there another solution?" Sweeney is ahead of his time. Of course, this begs the question:

FIGURE 6.5 Built in 1910 in response to the 1900 hurricane (described in Erik Larson's book *Isaac's Storm*), which killed at least 6,000 people, the famous Galveston, Texas, seawall is among the largest such walls in North America. As the beach gradually disappeared in front of the seawall, rocks were added to the base of the wall to prevent undermining. As the beach continued to steepen, a second row of rocks was put in place. In Hurricane Ike (2008), the wall prevented major wave overwash from the ocean, but much of the town flooded from the back side of the island. After the 1900 hurricane, the town raised its ground elevation 17 feet, using sand pumped in from the continental shelf, a unique and useful response to flooding that has not been carried out in the United States since then.

what does Florida's tourist industry do without beaches once the seawalls destroy them?

The Galveston, Texas, seawall, one of the mightiest in the nation, has held the line for more than 110 years and is still a functioning wall. But in Hurricane Ike (2008), much of the town was flooded by the seaward return of the storm surge onto the island from the back side of the island. This illustrates the point that the hard stabilization approach to protection from storms and sea level rise on this and most other barrier islands will ultimately require building seawalls around the entire island.

TABLE 6.4 Beachfront Mileage

Los Angeles County, CA	79 miles
Galveston, TX	32 miles
Ft. Lauderdale, FL	23 miles
Myrtle Beach, SC	60 miles
Atlantic City, NJ	8 miles
Fire Island, NY	32 miles

The beachfront mileage of various beach communities as claimed in their promotional literature. This provides an indication of the potential future cost of beach nourishment or seawalls as sea level rises. Open-ocean nourished beaches typically cost between $300 to $1,000 per linear foot and seawalls range in cost from $2,000 to $10,000.

Seawalls are common engineering structures and are a particular problem. Seawalls on an eroding shoreline always destroy the beach. Beach loss occurs because the seawall does not address the forces causing the erosion, so the beach continues narrowing until it's lost altogether. The time required for such destruction ranges from two or three years to two or three decades.

The cost of the hard stabilization response to sea level rise is high, both environmentally and monetarily. The cost estimates in appendix B should be viewed as approximations that can vary widely based on materials used and the nature of the ocean at a given location. Wave energy (height), tidal conditions, storm frequency, and likely storm intensity are important factors in the design of engineering structures.

Getting closer to the problem would be to multiply the hard stabilization costs by the actual beach mileage claimed by communities in their tourist brochures. The beach mileage of a few communities is shown in table 6.4.

Maintenance is also a significant part of the cost of holding a shoreline still. On shorelines all over the world there are remnants of wrecked coastal engineering structures reflecting a stormy past or, more likely, a past without maintenance. The 2016 U.S. National Park Service Coastal Adaptation Strategies Handbook suggests that breakwater maintenance would be on the order of $500 per foot per year. Maintenance of nourished beaches is particularly expensive, given that many such beaches

FIGURE 6.6 The iconic Breakers Hotel in Palm Beach, Florida, tries to fight beach erosion with replenishment and a seawall. The seawall has protected the building at the price of loss of the beach, an example of a global phenomenon. *Photo by Bookings.com.*

are effectively gone in a few years and require complete replacement on a periodic basis.

In a rising sea level, beach lifespans should shorten considerably, and the eventual result will be complete loss of the natural beach ecosystem. Seawalls probably are going to grow both in number and size in coming decades as beach nourishment becomes more costly, with ever-shorter lifespans (and ever-more-costly sand). Besides causing the loss of the entire beach ecosystem that provides the habitat for the food for shorebirds, seawalls will halt turtle nesting and sand crab feeding.

FIGURE 6.7 In spite of the steel seawall here on Galveston Island, Texas (west of the city), wave activity moved the shoreline back, a common event at seawalls around the world. Next, giant sandbags (Geotubes) were emplaced and covered with sand. Now the sand cover is disappearing, exposing the Geotubes, which will eventually fail, and the houses will fall in.

Obviously, the loss of the beach is a disaster for the tourism industry. Will we choose to protect the houses of a few at the expense of beaches used by many?

Multiplying the current costs of construction by shoreline mileage that may be "protected" by seawalls gives an idea of the immense costs of protecting coastal communities with seawalls. For example, construction of a five-mile seawall costing $2,000 per foot would cost some $52,800,000 (5,280 × 5 × $2,000 = $52,800,000). The numbers make the point that maintained hard stabilization on a broad front along many communities is not economically feasible in an age of rising seas.

In fifty years, beachfront communities will have two choices: hard stabilization or retreat. *Retreat from the shoreline almost certainly will ultimately prevail!*

THE ENVIRONMENTAL IMPACT
OF SURGING SEAS
· LIFE AT THE EDGE ·

Impact on Animals

Coastal animals face a number of problems from rising sea levels, including loss of beaches and complete inundation of habitats. It is important to note that beaches, left up to Nature, will always be present somewhere, but some will disappear simply because the sea rose up against a cliff or rock outcrop, or (often) bumped against a developed shoreline. Seawalls, houses, motels, septic tanks, roads, pipelines, and various other types of infrastructure along the shore will degrade or completely remove the beach habitat.

Sea level rise is not the only climate change–related hazard that life faces in and around the seas. Ocean acidification, the formation of carbonic acid from the CO_2 absorbed by the ocean, is already affecting the production of shell material by crustaceans and mollusks. Deoxygenation caused by nutrient pollution in the ocean is a particular threat to marine life. Areas of the ocean that are low in oxygen or altogether devoid of it (dead zones) are increasing. One dead zone is found in the Gulf of Mexico off the Mississippi Delta. And of course, ocean warming itself is responsible not only for sea level rise but is also wreaking havoc with many organisms, especially coral reefs in South Florida and around the world.

According to the Center for Biological Diversity (CBD), sea level rise threatens 233 species protected by the federal government in 23 coastal states. According to the CBD, the top 5 U.S. species most at risk from sea level rise are:

- Loggerhead sea turtle—due to beach narrowing, degradation, or loss
- Key deer—due to inundation of their island habitats
- Delmarva Peninsula fox squirrel—due to inundation from Chesapeake Bay
- Western and Eastern piping plover—due to beach narrowing, degradation, or loss
- Hawaiian monk seal—due to beach narrowing, degradation, or loss

Perhaps the greatest loss of animal life related to sea level rise is on the beaches, the main battleground between the land and the rising sea level. The impact there is primarily due to the human responses to the surging sea: construction of seawalls and nourishment of the beach (discussed in chapter 6). Seawalls eventually cause the loss of the beach habitat and all the organisms within it, including turtle nesting habitat. Since the fauna and flora of the beach make up the uppermost portion of the ecosystem of the inner continental shelf, the loss of critters in the beach sands impacts a large group of nearshore and onshore organisms.

Beach replenishment immediately kills almost everything on the beach, large or small, crawling or swimming. But as long as a beach remains, some of the animals will recover years later. How long the beach fauna and flora recovery will take is variable, depending on beach width and length, the quality of the nourishment sand, and whether the critter crawls, swims, or flies.

On every beach engineering project, there are two sites where beach nourishment has the most effect on living creatures. One is the beach environment itself, and the other is the offshore sea floor from which the sand is obtained. The first visual evidence of a dead beach is the lack of birds swooping back and forth and feeding in the swash zone. While

FIGURE 7.1 A nesting Least Tern on an Outer Banks beach in North Carolina. Obviously, such bird activity is much endangered by human activity, including driving on the beach and beach nourishment. *Photo by Sidney Maddock.*

nourishment is ongoing, when the sand is actually being pumped onto the beach, crowds of birds are "lined up" to participate in the big feed of dredged-up seafloor critters. However, that buffet is short-lived and the birds temporarily disappear along with most of the other life on the beach that has been nourished.

The ecosystem of the beach intertwines with the ecosystem of the inner continental shelf. Shore birds, surf fish, crabs, and some invertebrates comb the intertidal zone to feed and in turn are prey for organisms in deeper water. Shrimp fishermen have claimed a connection between poor catch off the Carolinas and recent beach nourishments, but this has not been independently verified.

Charles Peterson and associates at the University of North Carolina's Institute of Marine Science have noted that a nourished beach with coarse sediment containing abundant shells reduces the feeding opportunities for pompano fish that feed in the surf zone. A high concentration of suspended matter also causes problems for Pompano feeding. Sediment that is too shelly (a common problem with nourishment sand) causes armor-

FIGURE 7.2 Ghost crab at Smyrna Dunes Park in Florida. Beach nourishment kills such animals, and the species may not recover for several years. *Photo by Andrea Westmoreland, Flikr.*

ing of the beach sand as the shells become concentrated in a layer at the surface. This reduces the number of invertebrates, including the Donax clam, reduces feeding opportunities for sanderlings and other birds, and reduces the abundance of ghost crabs because burrowing becomes more difficult. Nourishment sand containing rock fragments the size of pebbles or cobbles makes turtle nesting difficult or impossible.

Just as sand placement on a beach is a death-dealing blow to beach fauna and flora, dredging the seafloor is also a killing process. Evidence of this is the aforementioned swarming of sea birds gobbling up newly arrived critters as fast as they appear at the end of the dredge pipe. In some cases, for example, off Myrtle Beach, South Carolina, the dredge virtually vacuums a large area of the seafloor, removing the uppermost few feet of sand and pumping it ashore. Everything goes and everything dies.

Perhaps a less biologically damaging source for beach material would be dredging sand from sand ridges on the inner continental shelf (where they exist), taking a deeper portion of the sand rather than vacuuming up widespread sand from the surface of the seafloor. Dredging deeper may limit the destruction of organisms to smaller areas, but, in these cases, a depression is left behind after the project is completed. One problem

FIGURE 7.3 Nourished beaches can provide nesting sites for turtles to lay their eggs, but the success of the nesting depends on the quality of the sand that was dredged up and placed on the beach. In addition, the sex of the baby turtles is determined by the temperature of the surrounding sand, which in turn may be affected by sand grain size. In this photo, sand is flying into the air as this green sea turtle digs her nest. All green sea turtles (*Chelonia mydas*) are on the endangered or threatened lists in different parts of the world. *Photo by Mark Sullivan, NOAA Fisheries, Permit No. 1013707.*

with these sand depressions was illustrated off Grand Isle, Louisiana. The dredge depression reduced the wave friction with the seafloor experienced by incoming waves and caused an erosion hot spot on the shoreline at the beach.

Impact on Plants

A study of coastal plant species by Garner et al. (2015) indicated that sea level rise was likely a greater hazard to the long-term survival of plant species along beaches than the anticipated global temperature or rainfall changes. As would be expected, inundation, erosion, flooding, salinization, and loss of habitat were the main causes identified in this central

FIGURE 7.4 The area outlined on the continental shelf off Nags Head, North Carolina, shows the portion of the sea floor that is permitted for dredging to obtain nourishment sand. This is also the area on the continental shelf within which all organisms living on or within the sediment will be killed by the dredging. *Diagram courtesy of Andy Coburn, Program for the Study of Developed Shorelines, Western Carolina University.*

coastal California study. These factors are most likely the main ones that impact coastal plants everywhere.

Another important factor that will determine the fate of coastal plant communities is habitat modification by human development. A 2016 U.S. Geological Survey report notes that because of shorefront development few native plant assemblages remain near Hawaii beaches. Those plant communities that do remain provide important habitats for shorebirds and various invertebrates. As sea level rises, those shorelines that have no development will allow some upslope movement of plant communities, and they will be protected. Where shoreline development exists, however, the plant communities are not likely to survive—the point being that in Hawaii, as well as along all other U.S. ocean shorelines, the fate of near-beach plants corresponds closely to nearshore development by humans.

There is one more important factor related to the natural existence of coastal plants. According to Feagin et al. (2005), erosion or shoreline retreat caused by sea level rise constrains dune plants to a very narrow area, which prevents the normal dune plant succession process that would occur in a wider zone. As a result, the late plants in the succession fail to appear, creating a problem for their dependent faunal species. This conclusion was based on a study of Galveston Island, Texas.

Impact on Salt Marshes

Coastal salt marshes are important ecologically and commercially. They are important habitats for various fish species, shellfish, and migratory waterfowl. More than 60 species of water birds feed or roost in marshes off Virginia. Salt marshes also act as nursery grounds for a variety of commercial fishery species, including shrimp. In addition, they act as a filter for pollutants coming in from the upland and are believed to reduce storm surges that move across them. In extreme southern Florida, mangroves sometimes take the place of the salt marshes and perform similar functions.

Coastal salt marshes keep pace with the rising sea level through a combination of migration in a landward direction and upward building. Marsh buildup might not occur, particularly if there is some sort of barrier like a bluff, a seawall, or buildings that block landward movement. Widespread development (buildings, roads, and other infrastructure) along salt marsh shorelines (which do provide a fine view of the water) will be particularly damaging to the marshes as they prevent the marshes from moving inland. Widely used and rarely prohibited, bulkheads and seawalls along estuarine shorelines are sure death sentences for salt marshes in a rising sea because they cannot migrate.

A fairly new type of infrastructure, the living shoreline, is being used widely to stabilize estuarine shorelines. Basically, these are some kind of wall, bulkhead, revetment, or breakwater, behind which is salt marsh grass. The marsh grass is planted between the wall and the marsh edge, which gives the structure the name *living shoreline*. However, *armored marsh shoreline* would be a better and more realistic descriptive term.

In a time of sea level rise, a natural marsh will erode on the outer edge and migrate landward on the inner edge, all the while producing and supplying nutrients to the surrounding waters. Some "living shorelines" are simply another form of seawalls. Ultimately, they disrupt the natural process and, on barrier islands, hamper the islands' ability to thin out and migrate during sea level rise. One important question is whether the marsh behind a living shoreline revetment continues to supply nutrients to the estuarine waters and to furnish habitats and nursery functions. Currently, living shorelines are approved almost without question, as though the term *living* makes it an environmentally sound shoreline stabilization approach. In many cases it is not.

A three-foot sea level rise is expected to greatly reduce the extent of salt marshes everywhere along U.S. shorelines unless concerted efforts are made to allow landward migration of the marshes by not allowing construction of artificial walls, "living" or otherwise, that impede migration. Removing the trappings of humans to clear the way for marsh migration would be a costly effort and probably would not be a high priority in a society responding in so many other ways to the rising sea. The eventual loss of salt marshes will be an ecologic and economic loss to the nation.

Impact on Farming

The impact of sea level rise on agriculture will be greatest on the U.S. Atlantic and Gulf coastal plains and on the Mississippi Delta. These environments are low lying, with gentle seaward slopes, making large areas susceptible to rising waters that lead to saltwater intrusion from below and nuisance flooding from above. Already, farm fields along the margins of Delaware Bay, Chesapeake Bay, and the Pamlico and Albemarle sounds have been abandoned because of salinization. Increasingly, farmlands of the Upper Sacramento Delta region of California now experience periodic flooding related to sea level rise. As these floodwaters subside, they leave behind salt that can foul the farmland for years.

Saltwater intrusion initially may push the lighter freshwater layer upward and cause plant roots to drown in fresh water. Abundant evidence

FIGURE 7.5 This large stand of dead trees near Stacy, North Carolina, is a reflection of rising sea level. The trees may not have been killed by the saltwater intrusion. Instead, it is likely that the rising sea pushed up the lighter freshwater lens, drowning the tree roots. In the foreground is a salt marsh.

for this can be seen in the large and small patches of dead trees of all species in the lowermost fringes of the coastal plains, especially along the shorelines of bays and lagoons behind barrier island chains. Trunks of dead trees are particularly noticeable in the Inner Banks and along the shorelines of the Pamlico and Albemarle sounds of North Carolina.

Simultaneously with sea level rise, climate change will affect coastal agriculture through changes in amounts of rainfall, especially during La Niña years when droughts are likely. Besides their usual effects on plant growth, drought conditions will also allow greater saltwater intrusion, as there is less groundwater keeping the salt water at bay. In addition, the U.S. Atlantic and Gulf coastal plains will be affected by crop-damaging winds and flooding from intensifying hurricanes, creating a quadruple whammy for coastal crops: nuisance flooding, more intense hurricanes, variable rainfall (drought and/or heavier rain events), and saltwater intrusion.

INUNDATED INFRASTRUCTURE

· IMPERILED ENERGY FACILITIES ·

According to a 2012 study published at Climate Central, 287 energy facilities in 22 coastal U.S. states occupy land at four feet or less above current sea level. The report identifies 130 natural gas, 96 electric, and 56 oil and gas facilities built on land four feet above current sea level. The total jumps to 328 facilities at just five feet above current sea level. Florida, California, New York, Texas, and New Jersey each have between 10 and 30 endangered sites, mainly for electricity production in the first three states and oil and gas production in the latter two. Louisiana stands out with 114 total energy facilities, including 96 natural gas facilities on land *less than two feet* above local high tide.

A number of nuclear power plants located along the coast to take advantage of the cooling ocean waters are threatened by sea level rise. A more immediate and wider-spread climate-related concern for these nuclear plants, in general, may be the warming of the waters the plants use for cooling purposes. Droughts and heat waves are causing temperature spikes in the water used to cool nuclear plants. In some states, overly hot cooling water or drought-related water shortages have resulted in a temporary shutdown of reactors. In August 2012, Millstone Nuclear Plant

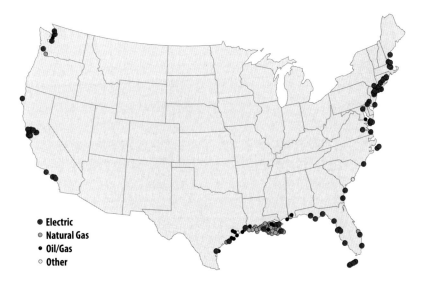

FIGURE 8.1 This U.S. map showing energy facilities less than 4 feet above local high tide illustrates their tremendous potential as a source for pollution from storm surges and sea level rise. *Map source: Climate Central.*

in Waterford, Connecticut, temporarily shut down a reactor when the temperature of the waters pumped in from Long Island Sound to be used for cooling purposes rose to 76 degrees. Record European heatwaves have also proven a challenge for the cooling of their nuclear plants, resulting in suspended operation and lessened energy output. It is ironic that global warming challenges the operation of nuclear power plants, an energy source that could be essential in decreasing greenhouse emissions.

Christina Nuñez, writing for *National Geographic*, identified 13 nuclear sites in the United States at risk due to sea level rise. Two of those plants, St. Lucie Nuclear Power Plant and Turkey Point Nuclear Generating Station, are in Florida. The eye of 1992's Hurricane Andrew passed directly over Turkey Point, a twin reactor plant near Homestead, Florida, destroying two water tanks used for fire protection and causing a long crack in a smokestack of one of the plant's oil-burning power units. One of the water tanks (a high tower tank) was blown down by the hurricane winds onto a 50,000-gallon water tank, disabling the fire protection system of the plant. David Lochbaum, director of the Nuclear Safety Project for the

FIGURE 8.2 The Turkey Point, Florida, nuclear reactors near Miami are at low elevation and therefore are very vulnerable to storms and rising seas. *Photo source: The Audubon Society.*

Union of Concerned Scientists, points out that the cracked smokestack could have fallen on the building that houses the backup diesel generators for the nuclear plant, and it was quite fortunate that did not happen because these were the only sources of AC power for Turkey Point for several days.

Seaside nuclear plants must prepare for a future of increased storm surge and severity of storms as the waters rise and warm. Prevention of nuclear disasters from storms could be challenging because coastal storms typically result in damage over a wide area, possibly making it difficult to get vital equipment and personnel to the site in the event of a storm-related disaster. This is exactly what happened at Turkey Point, a nuclear power plant with a worrisome history of unexpected problems. Lochbaum notes that while Turkey Point withstood the winds, Hurricane Andrew knocked out all off-site power to Turkey Point for five days, blocked roads to the plant with storm debris, knocked out off-site communications systems for four hours, and damaged the plant's radio system, leaving Turkey Point with one functioning handheld radio.

Notwithstanding those problems, Turkey Point was granted permission in early 2018 to build two more reactors, despite its location on a low-lying island on a southern Florida coast generally recognized as one of the areas most threatened by sea level rise. The current reactors are licensed to operate through 2032. Nuclear power plants are designed to last for decades, but in the coming decades, Florida will be experiencing significant sea level rise. Should Florida Power and Light (FPL) be allowed to bring more reactors into service on an island in a time of rising seas? In a letter to the Nuclear Regulatory Commission, Florida state senator José Javier Rodríguez called for increased safety measures or abandonment of the expansion of the reactors, pointing out that FPL's plans only take into account a one-foot rise in sea level when the area is likely facing a rise of at least three times that amount before the end of the century.

During high temperatures and drought in the summer of 2014, FPL was granted permission to operate their cooling canals at temperatures of 104 degrees, the hottest in the nation. The high temperatures in their cooling canal system, a 40-year-old canal two miles wide and five miles long, were also related to an up-rating of the reactors that increased power output by 15 percent to keep up with growing customer demand. The increased temperatures escalated evaporation and salt concentration in the closed system of canal waters that are drawn from a saline aquifer.

Unfortunately, readings show that Turkey Point is polluting a nearby drinking water well field with salt water. FPL estimates that, on average, about 600,000 pounds of salt (not salt water) seep into the groundwater daily. The saltwater plume has now migrated four miles westward and has pushed into mainland Florida, threatening a well field that supplies drinking water for more than three million South Floridians.

According to a 2015 study by David Chin, professor of civil and environmental engineering at the University of Miami, the cooling canal system contains water with a high concentration of tritium, a radioactive isotope that is commonly found in coolant water of nuclear plants. The presence of tritium in the aquifer confirms that Turkey Point's cooling canal is the source of the saltwater plume. In addition, the salt water is polluting Biscayne Bay, where elevated levels of tritium show that the canal's waters also are leaking into the bay. As of this writing, environmental groups are suing FPL, alleging they are in violation of the Clean Water Act. The state of Florida has given permission for FPL to charge

FIGURE 8.3 St. Lucie Nuclear Power Plant, Hutchinson Island, on the east coast of Florida, as viewed from the lagoon side. Since the ground elevation is probably 3 to 5 feet, this facility is in real danger from a major hurricane and sea level rise. Why build a nuclear power plant on a low-lying barrier island? The answer: plentiful cooling water. *Photo source: Nuclear Regulatory Commission.*

their customers for the estimated $176.4 million ten-year project to clean up the saltwater plume, passing the costs of cleanup on to the customers.

St. Lucie Nuclear Power Plant is north of Turkey Point, on Hutchinson Island, a barrier island near Port St. Lucie, Florida. According to the Nuclear Regulatory Commission, storm drains failed during a record rainstorm in 2014, flooding one of the reactors with 50,000 gallons of water and disabling core-cooling pumps. According to a Nuclear Regulatory Commission Notice of Violation against FPL's St. Lucie plant, if the nuclear reactor had shut down during the storm, the cooling pumps would have been submerged and recovery actions would have been necessary to cool the reactor. Damage to the reactor core could have occurred in 24 hours under that scenario. It is particularly concerning that inspections in 2009, 2010, and 2012 all failed to identify missing seals in electrical conduits that allowed floodwaters to enter the building, leading to disabled core-cooling pumps.

It is paramount that states take into consideration sea level rise when considering placement of nuclear plant construction. Seaside construction should be avoided because we are in the midst of historic sea level rise. Why construct nuclear power plants that could be flooded within the lifetime of the reactor? Accordingly, the expansion of Turkey Point's nuclear facilities would best be rejected. An underlying question that demands an answer: Should low-lying coastal Florida have *any* nuclear power plants in a time of rising sea level?

Another looming issue for coastal nuclear plants is the decision of when to decommission the plants and what to do with the nuclear waste. These nuclear plants, with a typical lifespan of 40 to 70 years, face decommissioning in the context of sea level rise and being further threatened by more coastal erosion and flooding.

Decommissioning nuclear power stations is a process that dismantles the plants to the point that they no longer require radiation protection measures. Decommissioning a plant can take decades and, according to Paul Genoad, director of policy development at the Nuclear Energy Institute, typically costs $500 million per unit. One particularly concerning challenge for coastal plants is that the spent nuclear fuel rods are typically stored on site, in part because no permanent repository for spent fuel exists in the United States. Initially, spent fuel rods must be submerged in pools of water to draw away heat to keep the rods at a safe temperature. After five years in a pool, the rods are cool enough that they can be separated and stored in dry casks. It is clear that decommissioning can't take place overnight and must be done before seawaters inundate the sites.

An obvious risk to the low-elevation energy infrastructure is flooding. Hurricane Sandy famously flooded a Consolidated Edison (Con Ed) power plant, plunging parts of Manhattan into darkness. The bright flash of an exploding substation that marked the beginning of the blackout was captured on video and can be found on YouTube. In the aftermath of the storm, Con Ed elevated vital equipment of the power plant, but the fact remains that the plant lies just 50 yards from the water. Like their nuclear sisters, conventional power plants are often built close to water, which is used for cooling purposes. A study published in the journal *Climate Change* (Rosenzweig et al. 2011) identified 20 power plants in New York City at risk of flooding because they are located at an elevation of 10 feet or less. Low-lying substations are also at risk and if a substation is flooded, then the power will go out. In addition, salt water can erode and inundate infrastructure, and saltwater intrusion leads to increased corrosion of the power infrastructure, which can also lead to power outages.

Wastewater and Storm Water Control

In a thorough evaluation of infrastructure at risk to sea level rise, published in the journal *Sustainability*, Beatriz Azevedo de Almeida and Ali Mostafavi (2016) point out that during storm events, when the power infrastructure is at risk of flooding, there is an increased demand for power used by stormwater drainage pumps, to prevent roads from flooding and for desalination of the water supply. This puts places like Miami and New Orleans at particular risk for increased flooding due to electrical failures during flood events. Energy demand will increase along the coast as the need for pumping and diverting storm-drain waters increases with sea level rise.

Especially vulnerable to inundation, wastewater plants, by design, are located at low-lying areas to minimize the need for pumping by taking advantage of gravity. During storms, flooded wastewater treatment plants sometimes release untreated waste into nearby streams and rivers. A 2014 report from the Center for American Progress reveals that close to 11 billion gallons of raw sewage ran into New York City streets, rivers, and coastal waters during Hurricane Sandy—enough to fill the Empire State Building 14 times! Hurricane Harvey (2017) shut down 40 wastewater treatment plants in the area of Houston, Texas, alone.

Much of the wastewater infrastructure in the United States is aging and in coastal areas is particularly at risk to sea-level-rise-related threats of flooding, rising water tables, and corrosion of pipes due to increased salinity in the water tables. When wastewater plants become overwhelmed with excess water, besides overflowing into streams, sewage can back up in pipes and flow into people's homes.

Sewage spills pose significant health risks. A 2006 study by Suzan Given and associates found that fecal contamination of ocean waters in Los Angeles and Orange counties alone cause as many as 1,479,200 gastrointestinal illnesses every year, with a public health cost of between $21 million and $51 million. In 2004, the Environmental Protection Agency (EPA) estimated that between 1.8 million and 3.5 million Americans contract illnesses each year from swimming in waters contaminated by sewer overflows.

FIGURE 8.4 The Hyperion Water Reclamation (wastewater) Plant, Los Angeles, California, like most sewage plants, is at a relatively low elevation so as to reduce pumping requirements to bring sewage into the facility. Thus, this facility and many others will be damaged by sea level rise long before higher elevation portions of the city are affected. With the destruction of the wastewater plant, more people will be affected by its loss than by actual flooding of the city. *Photo source: Google Earth.*

A 2018 study by Michelle Hummel and associates estimates that a three-foot rise in sea level would result in loss of wastewater treatment for 8.4 million Americans, including 1.5 million in the states of California, New York, and Virginia alone. This study shows that the number of people impacted by sea level rise due to loss of wastewater services could be five times as high as the number of people who are expected to experience direct flooding of homes or property. This means that municipalities need to take actions to elevate or relocate wastewater treatment plants to avoid their chronic flooding and the disruption of sewage treatment. The study points out that the plants are threatened not only by inundation from rising seas and higher storm surges, but they are also threatened by flooding from rising groundwater levels. Hummel et al. also found that 80 percent of the California treatment plants expected to flood in coming years are located in the San Francisco Bay Area.

Saltwater Intrusion

Worldwide, more than a billion people depend on coastal aquifers for their drinking water. A coastal aquifer is one in which the groundwater layer is in contact at its seaward end with salty ocean water. Normally along a coastal plain or barrier island coast, fresh water from rainfall on land will seep into the groundwater layer, gradually flow into the ocean, and keep the salt water at bay. But when this water is being used up before it reaches the sea, the flow is reversed and salt water flows inland.

Saltwater intrusion occurs when too much fresh water is extracted from the aquifer. This intrusion of salt water into the freshwater layer can destroy its use for drinking and agriculture plus myriad other uses, ranging from industrial utilization to landscaping to car washing. Saltwater encroachment can lead to abandonment of wells or the need to process groundwater through desalination plants. The problem is intensifying because of fast-growing coastal populations.

Saltwater intrusion and abandonment of water wells are occurring in diverse locations, including many towns on Cape Cod, Massachusetts, and the cities of Cape May, New Jersey; Hilton Head, South Carolina; Savannah, Georgia; Miami and Tampa, Florida; Los Angeles, California; as well as areas near the Delaware and Chesapeake bays. The intrusion of salt water into fresh water is a problem for almost every town along the coast of California and is exacerbated by chronic drought conditions in the state depleting the aquifers. At this point in time, sea level rise, slight as it may be compared to what will occur in coming years, is already an important contribution to saltwater intrusion, and it will become the most important component of such intrusion in a few decades.

Hilton Head Island, South Carolina, is a good example of the problem, as most of the wells on the resort island already have been eliminated as water sources because of salt penetration. Part of the predicament there is due to a huge groundwater cone of depression in the Upper Floridan aquifer around the nearby city of Savannah, Georgia—caused by pumping out water for that city's and Hilton Head's needs. This has reversed the flow in the aquifer and drawn salt water from Port Royal Sound toward Hilton Head. This intrusion has occurred simultaneously with a significant increase in the demand for water in the region caused by growth of industry and commerce in this popular tourist destination and by Savannah's switch from relying on river water to wells as a source of water supply.

Along the Pacific Coast, the problem is enhanced by a multiyear drought that has reduced the normal input of rainwater to the groundwater. This has accelerated saltwater intrusion, which has a particularly heavy impact on the rich agricultural area along the central California coast. On the Atlantic Coast, along Georgia and South Carolina, the saltwater intrusion is handled by restricting the gallons of water that communities can pump from the aquifer. In California, recycled wastewater is pumped into its aquifer in the hope of pushing back the advance of the saltwater front. In Hampton Roads, Virginia, treated wastewater is being pumped into its aquifer to minimize its depletion. The wastewater injection is intended also to reduce land subsidence, which in turn reduces the amount of sea level rise.

The loss of vegetation dependent upon fresh water has increased along shorelines where salt water has already intruded. Eventually, as sea level rises, salt marshes will replace the original vegetation. As mentioned in chapter 7, saltwater intrusion in the coastal zone is often manifested by stands of dead trees that are becoming increasingly common along the U.S. coastal plain. Rather than death due to salt water, the rising groundwater table may drown the tree roots before salt water reaches them.

Tidal flooding and storm surges can also be responsible for the contamination of the soils over which the salt water floods. The degree of salt contamination by this means will often depend upon the accompanying rainfall—the greater the rain, the less salt intrusion. In rural eastern Hyde County, North Carolina, near the town of Swan Quarter, you'll find a dike built specifically to protect agricultural fields from saltwater intrusion.

Desalination is often discussed as a solution to saltwater intrusion. However, this is a very costly source of drinking water, especially for large communities—at least two to three times more costly than normal groundwater production. But desalination is the wave of the future. It is a process that freshens saline water from the ocean as well as brackish water from bays and lagoons and wells. In the United States, there are more than 2,000 major desalination plants, mostly in California, Arizona, Texas, and Florida. Large desalination plants can produce in excess of 25 million gallons of uncontaminated water per day. The nation's largest ocean desalination plant, in San Diego, can produce 50 million gallons per day but even this amount won't supply all the water needs of this region.

The disadvantages of desalination are its high cost, energy hunger, and the problem of disposal of the potentially damaging super-saline water left over. The advantages of desalinized water are that in some cases it is a critical source of drinking water, it is drought-proof, and, in the use of ocean water, there is an infinite supply.

What to expect: restrictions on groundwater extraction in coastal areas to stall the saltwater intrusion, which will mean greater need for pumping water from inland areas and could also encourage increased saltwater desalination efforts. Increased desalination will result in greater energy demands occurring simultaneously with the need for increased pumping of water, another energy intensive activity; both processes will further strain our electric grids.

Transportation: May the Road Rise with You

Sea level rise and the increased intensity of storms expected with climate change put our nation's roads, rails, and ports at risk and will require significant funding to maintain and defend them against the coming floods.

In Kim Stanley Robinson's novel *New York 2140*, an example of the increasingly popular "Cli-Fi" genre of dystopian climate change–related stories, sudden sea level rise has disrupted worldwide trade by flooding port facilities worldwide. More than 80 percent of the world's freight is transported via ships. Sea level rise will threaten all the world's ports simultaneously, which means there will be a global effort to simultaneously raise the infrastructure of ports and employ storm-surge protections. This will be happening at the same time coastal communities will be fortifying their coast and raising or moving infrastructure. Ports will be prioritized as they are crucial to trade, but the costs of preparing ports for higher waters will be immense. For instance, damage to Louisiana's ports from Hurricane Katrina alone cost an estimated $1.7 billion to repair. Another challenge for ports is that we do not know exactly how high the seas will rise or how soon. Most likely, raising coastal infrastructure will be an ongoing project for centuries to come as the seas creep higher and higher.

What can we expect in terms of impact on transportation? As the waters rise, road bases and rail beds will be subjected to erosion, flooding, and eventual inundation. Traffic patterns will be altered as some roads flood, causing congestion elsewhere. Increased storm intensity will result in damage to infrastructure. Bridges, roads, and rail lines designed for twentieth-century conditions will buckle due to rising groundwater, failed flood-control systems, and the higher storm surges and greater-intensity storms that are expected with climate change.

Amtrak's busy Northeast Corridor route, which carries 12 million passengers each year between Boston and Washington, D.C., faces significant threats from sea level rise. According to an Amtrak climate change vulnerability assessment obtained by Bloomberg through a public records request, by mid-century, parts of the route face "continual inundation." Flooding, rising seas, and storm surge threaten to erode track and knock out signals that direct traffic, and substations and poles that provide electricity for trains are also threatened. The article notes that there is no alternative route. The assessment identified one severely threatened 10-mile stretch of track that runs perilously close to the Christina River in Delaware. The report calls moving the tracks "extreme," and suggests a solution of temporary flood barriers to be erected before rainstorms. On the West Coast, the Amtrak Cascades train hugs the Puget Sound as it winds its way to and through Seattle. In the future, trains will either be significantly rerouted on existing tracks or the coast-hugging tracks will have to be moved inland on newly purchased rights of way.

One obvious response to sea level rise is the raising of infrastructure. This is being done right now in some communities along the coast. Miami Beach is embarking on a $100 million project to raise streets, upgrade sewage connections, and install stormwater pumps to help it combat sea level rise. This is actually the start of a possible $500 million project that includes installation of as many as 80 pumps. The roads will be an average of 2 feet higher. Local residents fear that the raised roads will result in flooding on their property. This is a reasonable fear, as elevated roads can cause water to pool during flood events, ponding water and flooding nearby neighborhoods. Perhaps Miami Beach would be better off raising the entire community like Galveston, Texas, did (by 17 feet) in the early twentieth century after devastation from the 1900 hurricane. Ultimately, however, the waters will inundate Miami Beach. It's just a matter of time.

Raising coastal roads and streets is a temporary fix in many areas, as the seas will eventually rise and swallow coastal highways. It is a temporary fix, and it is not cheap. The higher you raise a road, the higher the costs. In the Florida Keys, estimates place the cost of elevating a 1.29-mile section of road in Big Pine Key at close to $9 million. Interestingly, the county hopes to collect and filter the runoff waters before pumping them into the ocean. In total, nearly half of the 300 miles of roads in Monroe County, Florida, home to the Florida Keys, are threatened by the sea level rise expected in the next 20 years. So, roads will be raised—but what about driveways and houses? Ultimately, in the Florida Keys, "resilience" is futile. The waters will come, and the Florida Keys will no longer be habitable.

So, it will take $9 million to raise up just over a mile of highway on the Keys, and there are around 150 miles of roads in Monroe County that are threatened. Examine the Keys on a map and you'll find this is just the tip of Florida, and, in terms of the costs of raising roads on all the U.S. coast, this is just the tip of the iceberg. New York City itself has over 500 miles of coast to protect. And all of this construction will be happening at once as communities prepare to be flooded by the rising seas. The costs are unfathomable. Ultimately, there will be retreat from vast portions of our coast.

A Future Full of Toxic Floods

Superfund sites are contaminated areas designated by the EPA for cleanup because they pose risks to human health or the environment. Hurricane Harvey flooded 13 Superfund sites in the Houston area, most of them related to the petroleum industry. Using EPA information, census data, and flood maps, the Associated Press (AP) identified 327 Superfund sites in areas prone to flooding or vulnerable to sea level rise. The AP found that nearly two million people, many of them low income and minorities, live within one mile of these sites. New Jersey and California lead the way with 35 sites each, followed by Florida with 33 sites and Texas with 24 sites. The EPA's 2014 Climate Adaptation Plan noted that prolonged flooding at these sites could cause extensive erosion and carry away contaminants as the waters recede, spreading the toxins.

In addition to Superfund sites, many active industrial sites with toxic chemicals are located in flood-prone areas. A *New York Times* study found more than 2,500 sites that handle toxic chemicals in flood-prone areas, including 1,400 that lie within areas designated by FEMA as having the highest risk of flooding. This list does not include Superfund sites, wastewater facilities, or chemical sites where the predominant risks are fire or explosion as opposed to toxic pollution. The *Times* noted that a Chevron Phillips plant in Baytown, Texas, built in a flood zone deemed to be at moderate risk of flooding, released 34,000 pounds of sodium hydroxide and 300 pounds of benzene, both highly toxic, during Hurricane Harvey. Despite lying in a moderate-risk flood zone, this was the third time in three years that heavy rains caused chemical leaks at the plant. The increased intensity of storm events shows that our flood mapping is inaccurate and obsolescent. What's needed are cleanup of Superfund sites that lie in flood-prone areas and, in order to avoid future toxic floods, taking sea level rise and changing storm-related flood risks into consideration when locating sites that handle toxic chemicals.

Coal ash, a waste product from coal-powered energy production, contains multiple toxins, including mercury, arsenic, lead, and radioactive materials. Rains from hurricanes Matthew (2016) and Florence (2018) caused coal ash spills in North Carolina. Reporter Sharon Lerner wrote that five coal ash sites owned and operated by Duke Energy spilled during Hurricane Florence. One of these spills during the record rains from Hurricane Florence turned a section of the Cape Fear River a sickly gray color.

Duke Energy pleaded guilty to nine criminal violations of the Clean Water Act from an earlier spill and agreed to pay out $102 million in fines. A 2015 EPA rule under the Clean Water Act required that any coal ash storage facility within five feet of a groundwater aquifer be closed. The EPA under the Trump administration subsequently extended the time coal companies have to close those ash ponds by 12 months, allowed states to suspend the monitoring of some groundwater near coal ash waste sites, and removed a requirement that only engineers can sign off on changes to coal ash ponds. Meanwhile, the coal industry is currently arguing in the courts in various states that the Clean Water Act does not apply to coal ash ponds or landfills.

Just how much of a threat is coal ash? To workers who cleaned up a coal ash spill in Tennessee, the threat has been deadly. In November 2018, a federal jury ruled in favor of workers who were sickened during cleanup of the nation's largest coal ash spill, the 2008 spill at the Tennessee Valley Authority's Kingston Fossil Fuel Power Plant. In the 10 years since the spill, some 30 workers have died and another 250 are sick or dying. Even some family members who were exposed to coal ash-tainted clothing are apparently sick. In a toxic tort lawsuit, the jury ruled against Jacobs Engineering, the contractor in charge of cleaning up the spill.

Unfortunately, many coal ash ponds and landfills are adjacent to waters, some along the coast. Sea level rise and the increased heavy rain events anticipated with climate change will continue to damage and destroy these toxic sites and pollute neighboring communities unless action is taken to relocate them or to improve the construction of the storage sites.

World Wide Web Underwater

Hurricanes and storms disrupt communications, making storm recovery all the more challenging. Hurricane Maria's devastating combination of rain, flooding, and winds destroyed not only the power infrastructure, but it also destroyed the telecommunications infrastructure, including cell phone towers. The devastation from the storm was all the more terrifying as family members outside of Puerto Rico found they could not check in with loved ones on the island. As frightening as the destruction of lines of communication might seem, it was, after all, only temporary. While we mainlanders haven't experienced a disruption in communication for as long as Puerto Rico did after Hurricane Maria, we have come to expect temporary communications disruptions following natural disasters. But climate change may prove a more long-standing threat to telecommunications and to the physical infrastructure of the internet in particular.

A University of Oregon and University of Wisconsin–Madison study used NOAA sea-level incursion projections and internet infrastructure deployment data and found that 4,067 miles of fiber conduit will be under water in the United States in the next 15 years. While standard internet infrastructure is weather- and water-resistant, it is not waterproof, and we can expect increased corrosion and interruption in service from sea level rise. In addition, the study found that 1,101 nodes will be surrounded by water in the next 15 years. These include points of presence, which typically house network interface equipment such as servers, routers, network switches, and collocation centers, which are hosting facilities where customers can rent space for servers and other computer hardware. The study identified the internet infrastructure of the cities of New York, Miami, and Seattle as the most vulnerable to the rising seas.

COASTAL CATASTROPHES

· CITIES ON THE BRINK ·

In January of 2016, General Electric (GE) announced its intention to move its headquarters from Connecticut to South Boston, Massachusetts—more specifically, to a flood-hazard area along Boston's inner harbor. Much of downtown Boston is on low-lying landfill formed mostly from sand and mud dredged from Boston Harbor. The land-making began in the eighteenth and nineteenth centuries when trash, mud from tidal flats, street sweepings (with tons of horse manure), and household coal ashes all were used. As recently as the twentieth century, trash was included in the landfill material. Although it is likely that in a few decades downtown Boston will be (temporarily) protected by a big seawall, why didn't GE, with all its technical expertise, relocate to an area that won't be part of the climate change problem in the future?

On the other side of the continent, Facebook, Google, and Cisco, among other Silicon Valley headquarters, are located at low elevations along the margins of San Francisco Bay. Just as in Boston, these iconic headquarters are built on fill dredged up from the adjacent bay floor, plus trash, earthquake rubble, and even sunken ships. The 430,000-square-foot complex of Facebook buildings is at particularly high risk, and the roads leading to the headquarters already face occasional nuisance flooding, with undoubtedly increased flooding episodes to come in future decades.

Here we have the smartest folks in the room—Facebook, Google, and Cisco in Silicon Valley and General Electric in Boston—building in locations that will in the short term require significant protections to avoid flooding (nuisance flooding or chronic flooding) or becoming isolated by nearby flooded streets, or in the long term will simply be flooded (chronic inundation). It is a clear indication that even in such intellectually heightened circles, sea level rise remains a largely academic subject: important, but not important enough for a major and globally significant institution to concern itself with.

One factor here leading to a contempt for Nature's processes is a *Too Big to Flood* mentality, a variation on the theme that *it can't happen to us.* This mentality must also be prevalent along the high-rise-lined shorelines of Florida. How else to explain the continued flourishing of the beachfront high-rise-condominium real estate market in the face of all the evidence of a rising sea? The assumption must be, even among those who recognize the problem, that the taxpayers will surely bail us out and protect us from Nature, won't they? In Florida, the situation has not been helped by state or national politicians who steadfastly refuse to believe in or act upon global climate change.

The Too Big to Flood mentality also applies along socioeconomic lines. It is assumed that exclusive beachfront developments will be protected if for no other reason than the huge political clout wielded by such wealthy communities. They are *Too Rich to Flood. Surely, as rich and important as we are, the government won't let the sea level rise touch us.* This is an absurd attitude, but it appears to be the state of affairs in several places.

In North Carolina, residents of the most exclusive and wealthy island, Figure Eight, led the battle to change the state's excellent anti-seawall laws. The same is true for exclusive Debidue Island in South Carolina, the residents of which fought the anti-seawall regulations of South Carolina. Clearly this Too Rich to Flood outlook results from contempt for the forces of Nature.

Then there is the thinking expressed by, among others, the mayor of Miami Beach: technology will come to the rescue, bringing a solution to damaging sea-level-rise impacts in the future.

In America's shorefront future, we must recognize that nothing is large enough to be worthy of taxpayer protection forever, and nothing is immune to the forces of nature. Nothing is Too Big to Flood and nothing is Too Rich to Flood, and future technology will not halt the rising sea in the foreseeable future.

The Outlook for American Cities at the Coast

Rising seas will affect different cities in different ways, mostly depending upon their elevation. As mentioned in chapter 5, the world's port cities, or at least the port facilities and surroundings, including docks and container freight yards, are all at the same elevation and hence are all facing the same risks. The docks all will need to be raised, which will become a costly accommodation for the changing sea level.

Of course, the magnitude of the impacts of higher sea level varies considerably. Miami, a city sitting at low elevation atop a very porous limestone, has virtually no real defense against a rising sea. It is a doomed city. Seawalls, levees, and dikes will not stop the floods, and massive water pumps won't help. Boston, with its low-elevation downtown area, can be defended eventually but only temporarily by seawalls. The fringes of New York are vulnerable but could be protected somewhat by walls. Some construction of walls (e.g., the Big U) is already underway. The flooding of subways in New York City by Hurricane Sandy in 2012 vividly illustrated the impact of a major storm surge. Houston, which does not sit in sight of the sea, is nonetheless vulnerable to climate change because of the expected intensification of future storms and storm surges. In 2018, David Ige, Hawaii's governor, in cooperation with the legislature, passed bills requiring that an analysis of sea level rise be included in all future environmental impact statements for development projects. Similarly, Honolulu Mayor Kirk Caldwell issued a directive for all city agencies to take action to develop land use policies, building codes, and hazard mitigation strategies to prevent the worst impacts of sea level rise.

New Orleans, like Miami, is a doomed city. As sea level rises, storm surges will be higher. At some point, the levees around New Orleans will again be topped in a massive way, reminiscent of the onslaught of Hurricane Katrina in 2005. On the West Coast, much of San Francisco is at

FIGURE 9.1 The Honolulu beachfront, like those of so many other coastal cities, has dozens of large buildings jammed up against the beach. Florida has hundreds of miles of high-rise-lined shorelines. In all cases, we can expect that the rising sea will cause loss of the beach, to be replaced by large seawalls. *Photo source: Garman, Creative Commons.*

"safe" elevations, but its fringes along San Francisco Bay, which are currently being heavily developed, are low and susceptible to flooding and the rising seas.

Now we are beginning to enter the era of legal battles over what to do with real estate on the beach, on land soon to be submerged as sea level rise progresses. In South Nags Head, North Carolina, six rental houses owned by one individual have been standing in the surf zone for years. They were once on row 2, but erosion has made them row 1. The Town of Nags Head proposed to remove the houses as a nuisance because they prevented any sort of recreational use of the beach. However, the state required the town to purchase the buildings before removing them, which they did, eventually, for $1.5 million. Remaining abandoned houses are still in peril, as the town has no funds to pay for their removal. They await the next big storm. Waiting for the storm to remove buildings will leave debris on the beaches for miles around. The South Nags Head problem may be repeated in larger coastal cities, such as Wilmington, North Carolina, and Norfolk, Virginia.

There is also a large difference in the preparedness of different cities for the coming sea level rise. New York, traumatized by Hurricane Sandy, leads the way in beginning to take real steps in the right direction. Tampa/St. Petersburg is huffing and puffing but not really facing up to the prospect of a three- to six-foot sea level rise. Charleston, South Carolina, and Norfolk, Virginia, both low-lying cities already suffering from frequent tidal flooding, have taken important steps to reduce the impact of sea level rise. Wilmington, North Carolina, is in an early planning stage. However, the state of North Carolina is practically ignoring sea level rise and passed a law in 2012 that actually blocks state agencies from using predictive methods to determine and respond to the impact of sea level rise. This has become perhaps the most backward state sea-level-rise program in the country.

In a 2018 *Scientific American* article, Jen Schwartz describes a unique method to use for retreat from the rising seas. New Jersey has recently formed a program known as the "Blue Acres Buyout Program" directed by Fawn McGee of the New Jersey Department of Climate Protection. Blue Acres has three major goals, which are somewhat similar to the LA SAFE program in Louisiana discussed in chapter 1. The first is to move people out of the way of tidal flooding and of flooding by storm surges. Second, the purchased houses are removed or demolished, and the newly restored land becomes available for public use. The third goal is to restore the ecology to its natural state so the land will act as a sponge absorbing rainfall and tidal flooding waters, reducing the flood risk to adjacent communities.

Hampton, Virginia, is following a similar path. Patricia Sullivan, writing in the *Washington Post*, notes that Hampton has purchased 18 houses that had been frequently flooded, "razed them, and turned the land into a wildlife and native plant marsh with a recreational boardwalk and trail." The city has also elevated 11 houses (with a combination of FEMA and city money) and plans to elevate 26 more.

A 2017 research report by Climate Central listed the top 25 coastal cities and their populations at risk of flooding within FEMA's designated

TABLE 9.1 Cities Most Vulnerable to Flooding by 2050

	City	Population at Risk
1.	New York, New York	42,600
2.	Hialeah, Florida	20,400
3.	Miami, Florida	15,400
4.	Fort Lauderdale, Florida	12,700
5.	Pembroke Pines, Florida	12,000
6.	Coral Springs, Florida	11,900
7.	Miramar, Florida	10,000
8.	St. Petersburg, Florida	9,100
9.	Davie, Florida	9,000
10.	Miami Beach, Florida	8,700
11.	Charleston, South Carolina	8,300
12.	Pompano Beach, Florida	8,000
13.	Sunrise, Florida	7,900
14.	Hollywood, Florida	7,600
15.	Miami Gardens, Florida	7,200
16.	Norfolk, Virginia	6,600
17.	Lauderhill, Florida	6,600
18.	Cape Coral, Florida	6,600
19.	Boston, Massachusetts	6,200
20.	Tamarac, Florida	6,000
21.	Virginia Beach, Virginia	5,800
22.	Tampa, Florida	5,700
23.	Fountainebleau, Florida	5,600
24.	Margaret, Florida	5,300
25.	Kendale Lakes, Florida	5,300

Cities and their threatened populations that are most vulnerable to coastal flooding by 2050. This is based on the population expected to be living at or below FEMA's 100-year flood level projected to 2050 for sea level rise.

SOURCE: Kulp et al., 2017.

100-year floodplain at the projected sea level in the year 2050. New York City has by far the largest population at risk (426,000). The extraordinary risk of Florida communities to flooding and sea level rise is indicated by the fact that out of 25 communities, only 5 are *not* in Florida (New York, Charleston, Norfolk, Boston, and Virginia Beach).

A Sampling of Cities

Seattle

The West Coast is dramatically different from the East Coast in terms of potential impact of sea level rise. Both the East and Gulf coasts are fronted with a broad, low-elevation, gently sloping coastal plain. From the South Shore of Long Island to the Mexican border is an almost unbroken 3,000-mile-long chain of eroding barrier islands. However, the West Coast—the leading edge of the North American tectonic plate (discussed in chapter 6)—has higher beachfront elevations, steeper slopes, and more irregular topographic features. Commonly the shoreline is at least partially lined by hard rock. The West Coast local governments anticipate a two-foot sea level rise by 2100, which is conservative.

An interesting study by Buchanan et al. indicates that by 2050, "cities such as Seattle, that have little history of severe flooding, are likely to experience . . . an uptick in the number of severe or even historically unprecedented floods" (Zandonella 2017). This is true because in this period of quiet storms, buildings have rushed to the shore where they should not have. In contrast to Miami and other Florida coastal cities, the exposure of Seattle neighborhoods to the rising seas is generally restricted to a narrow band of neighborhoods immediately next to the shoreline.

Much of Seattle's Puget Sound shoreline is backed by bluffs that, under rainy conditions or during winter storms with high waves, have a tendency to fail (slump). Sometimes the slumps carry houses to the beach. Seawalls and bulkheads, very common along Seattle's shorelines, are often intended to hold back bluffs or were placed to protect buildings atop the bluff.

Unfortunately, erosion of the bluffs is the primary source of sand for the beaches, so the walls and bulkheads add to the widespread narrowing of beaches. One indication of this beach loss is the pile of rocks (rip rap) placed at the base of seawalls, meant to preserve the wall. As sea level rises, however, storm waves will do extensive damage to the walls, and loss of individual homes will ensue.

Industrial Harbor Island and residential South Park and Georgetown, the drainage basins for Thornton and Longfellow creeks, plus all port facilities, are the areas most threatened by sea level rise in the Seattle region. Flooding of homes along Lake Union and Lake Washington will depend on the height of the sea relative to the Ballard Locks. Unless the sea rise is greater than six feet, the locks should not be overtopped (except by storms). However, low sections of State Route 99 and State Route 522 are already vulnerable to flooding.

San Francisco Bay

San Francisco Bay is expanding, and so is the city of San Francisco. The city is expanding to accommodate an increasing population, and the bay is expanding to accommodate more water because the sea level is rising. In the not-too-distant future, the expanding city and the expanding bay are bound to collide, to the detriment of the city. Not the least of the problems of San Francisco is that 80 percent of the wastewater (sewage) plants in the state of California that are expected to flood are along the margins of the bay.

Including some of the Silicon Valley headquarters, nearly $100 billion in commercial and residential property around the Bay Area is at risk from sea level rise and from flooding in severe storms. Soon to be added to that number, $21 billion in new development is planned for areas at risk. Among the more questionable developments to be built in areas susceptible to flooding and eroding shorelines are the $1 billion arena (Chase Center) for the Golden State Warriors basketball team and the proposed 8,000 homes (and three hotels) in a development on Treasure Island next to the Golden Gate Bridge.

Urban growth will take advantage of four large plots of land along the shore of the bay, including Treasure Island, Mission Park, Pier 70, and Candlestick Point. The retreat option is not on the table—not yet—but at least they intend to develop in full recognition of the coming hazard. The city of San Francisco requires developers to take into account the projections of sea level rise up to the year 2100 and even beyond. The city baseline has been set at a 66-inch sea level rise, by 2100, that is deemed to be in the *unlikely but possible* category by the National Research Council. With this move, the city of San Francisco sets an excellent example for the rest of the nation. They could have chosen instead the more conservative route of what is considered the "most likely" category of sea level rise—36 inches.

San Francisco's 66-inch projected sea level rise in the next 85 years contrasts sharply with the 8-inch projection in the next 30 years that the state of North Carolina currently uses in planning for the rising seas. This 8-inch number reflects the essentially nonexistent coastal management program in the state.

Southern California

The Southern California coast has a number of towns with a high possibility of damage from sea level rise to buildings as well as to highways and other infrastructure built close to the shoreline. Like other portions of the American West Coast shoreline, the geology and physical nature of the shoreline here vary widely. Flooding of large inland areas, like so much of the East Coast, is not going to be a problem for the immediate future here, although Long Beach and the Port of Long Beach can expect flooding for several blocks inland. Communities facing damage from the rising seas include Point Loma, Laguna Beach, La Jolla, Del Mar, Oceanside, Malibu, and Santa Barbara. Malibu has a particularly spectacular lineup of houses essentially perched on the upper beach that are vulnerable in the extreme to the next big storm. Los Angeles has two power plants, two wastewater treatment plants, various waste disposal sites, and the port, all of which are in danger of being inundated by sea level rise and associated storms.

Anne Mulkern of *E&E News*, reported in late 2018 that the California Coastal Commission plans on releasing a guide in 2019 for dealing with rising seas in residential areas. The report states that protecting shorelines with armoring is an option, but it wisely discourages the use of sea walls because they will make sandy beaches disappear. The guide will apparently have a section on managed retreat, including suggestions on how to implement managed retreat, with the objective to preserve the state's beaches for the public. However, at least one California community, the town of Del Mar, has vociferously voiced their opposition to the idea of retreat. In 2018 they passed an ordinance promising to fight any effort to retreat.

Houston

Michael Kimmelman, architecture critic for the *New York Times*, put it this way: "Houston was founded in a swamp by a pair of real estate swindlers from New York [the Allen Brothers] selling faraway investors a bill of goods about the Promised Land [sounds like South Florida]. When oil turned up under the muck, they seemed like prophets. Today, the giant city's sprawl of highways and single-family homes represents a postwar vision of the American dream." It is the home of the space program, the oil capital of the United States, and a center of high-level medicine.

But it was a dream pursued without planning or zoning. It was an "anything goes" atmosphere with zoning controlled by the development industry with little concern for risks from storms (as Hurricane Harvey strikingly demonstrated in 2017). Houston embraced a vision of an unfettered America.

We have so much to learn from Houston's undisciplined approach to development. Homes were built in areas known to flood frequently, marshes that might have buffered or absorbed some flooding were paved over, and canals intended to direct waters away from the city were lined with buildings. Most stunning of all, houses were built within emergency floodwater retention reservoirs, allegedly without informing inhabitants that their homes were in a potential lake! In 1992, engineer Charles Glen Crocker wrote a series of letters to a variety of county officials pointing out that he had learned from the Corps of Engineers that the planned Cinco Ranch and Kelliwood subdivisions were within the lake that would form in the Barker reservoir during a heavy rain (Drew 2018).

Instead of getting the thanks he deserved, his credentials were questioned, and he was criticized for holding back development in the county. In Hurricane Harvey, over 5,000 households flooded in the reservoir, and so it went in the Wild West. More on the amazing reservoir development in chapter 10.

Many residents of Houston had experience with floods long before Harvey hit. As many as 250,000 people from New Orleans fled to Houston seeking shelter after Hurricane Katrina in 2005. More than 25,000 of

FIGURE 9.2 The Motiva oil refinery, Port Arthur, Texas, was partially destroyed in Hurricane Harvey (2017). In this photo, oil appears to coat most of the surfaces. The refinery's low elevation makes it very susceptible to future sea level rise and storm surges. *Photo source: Google Earth.*

the New Orleans refugees were still in Houston when Hurricane Harvey hit, and many were likely affected by flooding in this city, too.

In January 2018, the city announced that there was a new, ambitious master plan for downtown. According to city officials, the emphasis in the new design is on sustainable and resilient development, presumably utilizing lessons learned in Harvey. Among the lessons was the value of local and regional planning!

Three 500-year floods in Houston over three years' time show that something has changed (and also demonstrate the futility of FEMA's risk maps in an age of climate change and sea level rise). And the potential for more extreme events is high, ever more so as the sea rises. A worse storm is possible, one that comes up from Galveston Bay and up the Ship Channel to destroy oil refineries and create a wave of pollution pushing into the city's outskirts.

One of the biggest fears among Houston's civil defense and emergency officials is that the numerous polluted Superfund sites and petrochemical plants along the Houston Ship Channel will flood in a future storm surge. Under the right circumstances, a deadly soup of chemical pollutants could flood into the city. To offset this possible flood, a $12 billion Ike Dike (Ike refers to Hurricane Ike, which struck the area in 2008) is proposed on Galveston Island to prevent a storm surge from coming up the Houston Ship Channel and to protect development on Galveston Island. The plan is for a long seawall along all of Galveston Island and the adjacent Bolivar Peninsula, with a tidal gate across the inlet separating the two. This would, in a few years, take out the beach on the unseawalled portion of the island permanently. Planning for the Ike Dike is well underway. Better to spend the money cleaning up the areas along the Houston Ship Channel and moving, raising, strengthening, or demolishing endangered houses and buildings, and improving the drainage system in Houston in preparation for the next catastrophic rain.

According to the *Houston Post*, the designers of the Ike Dike have not considered the impact of a catastrophic rain like Hurricane Harvey's rainfall in Houston (an unbelievable total of 52 inches) or the harm to businesses that a storm surge would cause. Add to that the loss of the island's recreational beach.

The Ike Dike is a 19th-century solution for a 21st-century problem, with not only a high initial price but also assured long-term high maintenance costs. The Ike Dike is not the solution needed in a time of rising seas. It is a solution that assumes we can hold the shoreline in place. Ultimately, the Ike Dike will lead to denser development on Galveston Island, "protected" behind the seawall. A new disaster is forthcoming—perhaps like the 1900 hurricane that devastated Galveston and was memorialized in Erik Larson's book *Isaac's Storm*.

It is not clear where the future will take Houston. Buying out houses and returning green spaces to absorb water will be one approach. Regional planning to bring community planning together will be one step, but it's not there yet. Sea level rise will require changing the downtown into a safer one, a most difficult and probably impossible task. Many Houston dwellers on dangerous ground will benefit by moving to higher ground—out of town.

Some things never change. In Friendswood, Texas, a town of 40,000 people just southeast of Houston, town officials relied on a 20-year-old flood insurance map that was both out of date and invalid for the purpose of allowing 300 homeowners to rebuild at ground level. The flood elevations on the old maps are well below the Harvey flood level. The excuse for the rebuilding of houses at elevations that were flooded during Harvey was the need to help some struggling people stay because they couldn't come up with the money to raise their houses. So the 300 owners will remain at a very precarious elevation, and since the path was cleared for the 300 owners, hundreds of others will be allowed to follow. If county officials believe that they are helping these 300 people, how will they respond when those houses are flooded again? Why not give them some financial aid to raise or move their houses?

Tampa/St. Petersburg

In Tampa, segments of thirty-one local roads will flood in a storm or with sea level rise of 18 inches, blocking access to homes and businesses and complicating escape from them in storms. These include the airport access and Campbell Causeway access north and south. Thirty parks are at risk (better to flood parks than buildings and roads), as are many properties along the margin of Tampa Bay, McKay Bay, the Bypass Canal, and the Hillsborough River. Tampa General Hospital on Davis Island, a landfill, has generators and safe floors, but bridges to the island would flood in a storm. A map of Pinellas County indicates it will be split in two by a three-foot sea level rise, forming two islands. One island will be downtown St. Petersburg and the other will be Clearwater.

Rounding the numbers, the Tampa Bay Regional Council accepts NOAA's projections of sea level rise to the year 2100. The high projection is seven feet, the intermediate high is four feet, the intermediate low is two feet, and the low is one foot.

The Cost of Doing Nothing about the Rising Sea is outlined in a report of the same name by the Tampa Bay Regional Planning Council. "Rising sea levels may submerge many areas of the Tampa Bay region by 2060. While there are varying estimates, one worst-case possibility suggests that sea levels may rise as much as three feet over the next 40 years. Such an increase will cause thousands of properties to be flooded, thousands of jobs to be lost, lost taxes and lost income from tourism as well as the loss of much of the region's barrier islands." Thousands will join the stream of Florida climate refugees heading north.

TABLE 9.2 Impacts from Inundation of a Three-Foot Rise:

Direct impacts	
Value of submerged residential property	$14.9 billion
Value of submerged commercial property	$1.3 billion
Property tax loss	$5.4 billion
Job losses in submerged areas	17,184
Indirect impacts	
Gross Regional Product Loss	$2.0 billion
Non-Tourism job loss—Regional product loss	$4.4 billion
Tourism job loss—Regional product loss	$1.9 billion

SOURCE: Tampa Bay Regional Council, 2017

Why the Tampa Bay Regional Planning Council has chosen the time frame of 40 years is puzzling. New York City and most other communities look to the year 2100 and even well beyond for their planning purposes.

The Council framed sea level rise in seven different contexts or impacts, a useful approach for any region likely to be impacted by sea level rise and an indication of the complexity of determining potential economic impact of the rising seas.

The analysis by the Regional Planning Council concerns inundation (flooding) from sea level rise in 40 years and *not* the impact of storms. However, it is statistically likely that the Tampa Bay region will suffer significant damage from big storms within the next 40 years. And as the ocean warms, the probability increases that the storms will be record breakers. Each increment of sea level rise in coming years will increase the relative height of storm surges and of course increase the amount of damage.

According to the *Washington Post*, "A Boston firm that analyzes potential catastrophic damage reported that the Tampa Bay region would lose $175 billion in a storm the size of Hurricane Katrina. *A World Bank study called Tampa Bay one of the 10 most at-risk areas on the globe.* Yet the Bay area—greater Tampa, St. Petersburg, and Clearwater—has barely begun to assess the rate of sea level rise and address its effects. Its slow response to a major threat is a case study in how American cities reluctantly prepare for the worst, even though signs of impacts from climate change abound all around."

"People who want to live on the waterfront will always live on the waterfront," a local citizen, oceanographer Mark Luther, said, in a reference to the rich, in an interview in the *Washington Post* article. "Every house on my street that sold within the past 10 years, they've knocked it down and built a 10,000- or 12,000-square-foot mini-mansion on top of it."

To sum up the Tampa Bay regional effort to do something about the coming sea level rise: *Lots of Smoke but Little Fire.*

New Orleans

"There is a time and a place for politics, but this is not it," Louisiana Governor Bobby Jindal said. "It is therefore with disappointment that I read of the White House [Obama administration] plans to make this visit part of a tour for your climate change agenda. Although I understand that your emphasis in New Orleans will—rightly—be on economic development, the temptation to stray into climate change politics should be resisted."

This statement by the then-governor of Louisiana says a great deal about the future of Louisiana in a rising sea and explains why the current coastal management and preparation for sea level rise are based on very conservative estimates of the magnitude of the rise. The idea that global climate change is a political concept is most distressing and very wrong. It's science, not politics.

The story of Hurricane Katrina (August 2005) and its impact on New Orleans provides an example of the inadequate response of a major city to major flooding. During the storm, the levees failed and much of the city flooded, although the business district and the main tourist hot spots were spared. Pumps, intended to drain floodwaters, failed. Probably not helping matters were the 46 tons of Mardi Gras beads that have recently been discovered in the clogged street drainage basins.

In 2000, the population of New Orleans was about 485,000. According to the Data Center, the poststorm population in July 2008 was around 230,000, a loss of nearly half the prestorm population. By July 2015, poststorm confidence grew and, with the city's colorful history beckoning, the city population was back to 80 percent of what it was in 2000.

The fact is, a storm greater than a 100-year event could top the New Orleans levees again and drown much of the city. It would depend on the direction the storm is moving, the velocity it is moving (the slower, the

FIGURE 9.3 A flooded downtown street in New Orleans after the passage of Hurricane Katrina (2005). As seas continue to rise, the likelihood of an important hurricane flooding the city again is much increased. The long-term future of New Orleans is grim. *Photo source: FEMA.*

worse), the wind strength, and the storm size. It *will* happen someday. Meanwhile, the rising sea level will increase the probability of major hurricane flooding.

Sea level is rising very rapidly around New Orleans and the Mississippi Delta because:

- The sinking land caused by the extraction of oil and gas (and water) is probably the most important reason for the very rapid local sea level rise.
- The muddy sediment on the delta naturally compacts, causing the land to sink (and the sea level to rise).
- Like much of South Florida, New Orleans was built on swamps that were drained, causing the land to sink rapidly.
- Adding to the resulting sea level rise, the Corps of Engineers built dams on the Missouri River tributary to the Mississippi. These dams trap sediment and rob the delta of the sediment that used to build up the surface.

- The Corps of Engineers built levees along the river outlet channels across the delta leading to the Gulf of Mexico. As a result, the remaining sediment in the river water was channeled straight out to the Gulf, contributing to a biological dead zone offshore, rather than spreading out and being deposited on the delta.

One plan to slow down sea level rise is to divert most of the Mississippi sediment-laden waters out onto the delta surface. Torbjörn Törnqvist, a geologist at Tulane University, notes that "if sea levels start to ramp up, all these [sediment] diversions are going to do is delay the point of no return basically by a couple decades. It will be increasingly clear that the city is not going to survive. Then you will get into issues like, what are we going to do? Are we going to relocate the city? How that's going to play out, it's hard to predict. What's really critical is what kind of action will be taken within the next decades" (Magill 2015).

Miami

The Organisation for Economic Co-operation and Development (OECD) compared the relative property damage risk exposure of the world's top ten largest port cities to flooding from rising seas and storm surges. Miami topped the list.

Part of the reason for the lofty position on the OECD list is that Miami and Miami Beach, parts of Fort Lauderdale, and much of South Florida are underlain by a very porous and permeable limestone known as the "Miami Limestone," sometimes called the "Miami Oolite." The rising sea will simply rise through this limestone and flood the city, seawalls or no seawalls. In effect, Miami is like the floating villages in Cambodia, Thailand, Myanmar, and Indonesia, made up of houses on stilts.

Already, nuisance flooding is affecting low areas. Publicity has shone a particularly strong light on Miami Beach with photos showing cars plowing through flooded streets. Such scenes, spectacular as they may be, fail to illustrate the smell of backed-up sewers, a product of the floodwaters. Elsewhere in the city of Miami and especially in the rest of Miami-Dade County, foul odors are becoming more common as septic tanks by the thousands are failing. The failure is occurring because soils in septic fields

FIGURE 9.4 The Miami Worldcenter, an ongoing billion-dollar development, is a symbol of the developers' and city officials' belief in a long-term high and dry future. But the city sits atop a porous permeable limestone that will allow free access of the rising sea into the city. The Miami limestone means that the city is virtually standing in water, like ancient cities with homes on pilings in lakes. Miami is a doomed city within this century. *Photo source: Miami Worldcenter Associates.*

are wet due to a rising groundwater table being pushed up by sea level rise. Wet soils cannot absorb the effluent, which then seeps to the surface.

The city of Miami Beach is raising some streets to combat nuisance flooding, but as experience has shown on the Inner Banks of North Carolina, raised roads act as dams in storm surges and lead to increased flooding. Meanwhile, construction of new buildings in Miami continues unabated, and so far, Miami exhibits a lack of political will to even recognize the gravity of the problems that rising seas bring.

The following list summarizes the likely impact of sea level rise for Miami:

- Chronic and eventual complete inundation from sea level rise aided by the permeable limestone
- Increased height of storm surges due to both storm intensification and sea level rise

- Increased rate of failure of storm gates across drainage canals due to sea level rise and extreme rainfalls
- Salinization of the water supply
- Eventual massive relocation of the citizenry of Miami to higher ground, most likely out of Florida

There is no way that the Miami community is not aware of the sea-level-rise problem, thanks to Hal Wanless, chair of the Geology Department at the University of Miami. Wanless beats the drum of sea level awareness at every available opportunity and organizes seminars concerning the future of the embattled city. He and his science colleagues offer these no-cost seminars to the public—all who care to come through the door.

Recently, South Florida newspapers have banded together in an effort to increase awareness of the extreme hazard that this region faces from the rising sea. In a *Sinking Cities* segment, PBS listed the various problems facing the city but offered no realistic solutions and failed to include Hal Wanless among its guests. A problem is that so far, long-term solutions have not been proposed. In fact, in all reality, there may be no long-term solutions to the salvation of South Florida communities.

Charleston

Charleston has long faced the possibility of flooding, both of a tidal nature and from hurricane storm surges. Because the entire Charleston Peninsula area (the downtown) is flat and at low elevation, it already has a flooding problem, and higher seas will only exacerbate the problem. Like in another small and low-lying city, Norfolk, Virginia, the rising sea has received much attention from city officials. Charleston's sea-level-rise strategy assumes a rise of 18 to 30 inches within the next 50 years. The city recently passed a new regulation requiring the first floor of new construction to be built at least one foot above a designated base flood elevation. Also, it strongly enforces building codes and various other regulations designed to respond to flooding.

The city has spent or set aside $237 million to support drainage projects. These consist of a number of 10-foot- and 12-foot-diameter concrete pipes with pumping stations and discharge pipes leading to a nearby estuary. Recently, at a very small cost of less than half a million dollars, the city has installed check valves in their drainage pipes. The pipes let rainwater out but keep the sunny day floods at bay. So far it seems to have improved the situation considerably. As for future plans, the city is seeking $2 billion from state and federal coffers to extensively floodproof the city for decades to come.

Different solutions to the flooding problem will be needed in different parts of the city. It is almost certain that eventually some seawalls will be needed to protect the city center with all its historic buildings. A new seawall has just been built at the Battery, the southern tip of the city, from where Confederate artillery fired the first shots (although a possible urban legend says that Citadel cadets had earlier fired on a federal merchant ship) of the Civil War. It is an absolute certainty that Charleston and probably Norfolk and many other small coastal cities will not survive much beyond this century without becoming walled cities.

When Charleston is in real trouble with a big sea level rise, perhaps in 50 to 75 years, the city will be competing for federal funding with such major municipalities as New York, Boston, Miami, St. Petersburg, Houston, and San Francisco. The competition for support will be intense and politicized. The point is that taking early action to defend against a future rise before other cities start clamoring for money makes sense. And Charleston seems to be doing just that.

Norfolk

Norfolk, Virginia, the home of the world's largest naval base, seems ahead of the curve in recognizing its extensive vulnerability and is actively planning for a future life with a rising sea. Norfolk faces some of the fastest rates of sea level rise in the nation and should be applauded for recognizing that planning for sea level rise must include retreat from vulnerable flood-prone areas. Norfolk will be among the first U.S. cities to reinvent themselves.

The city has zoned itself to encourage development in safe areas and recognizes four color-coded zones. Green and purple are for relatively safe areas in this low-lying city, where the city plans to focus future growth. The red zone, comprised mainly of downtown Norfolk and the naval base, is a densely populated area in need of protection. Most impressive in terms of forward thinking, the yellow zone consists of areas the city cannot afford to protect from floods and where "innovation, adaptation or retreat will take the place of hardened protections like floodwalls or levees." Nicholas Kusnetz, writing for *InsideClimate News* blog, notes that the city has plans to tear down low-income housing in the red zone in favor of mixed-income development. Michelle Cook, president of Tidewater Gardens Management Corporation, one of the housing developments slated for destruction, fears that the city will not be able to accommodate all of the residents in the new mixed development.

New York

The assumption is made by New York City planners that the city will be affected by between 18 and 50 inches of sea level rise by the year 2100 (75 inches is considered a possibility) and that sea level rise is locked in for centuries to come.

Parts of New York City, like parts of Boston, have been created by landfill, which makes them low, vulnerable areas. Most of the damage from Hurricane Sandy in 2012 occurred in low-lying landfill areas at the lower tip of Manhattan. Brooklyn and Queens and the margins of Staten Island also have large low-lying areas so are also susceptible to flooding. The city's shoreline measures an impressive 520 miles in total.

Thanks to the disaster wrought by Hurricane Sandy, New York now is probably far ahead of most other communities in planning to prepare for sea level rise. Hurricane Sandy affected roughly 88,700 buildings containing 300,000 homes and 23,000 businesses. Mayor Michael Bloomberg announced in 2013 a $19.5 billion, multi-decade plan to protect the city from big storms and sea level rise. The plan called for levees, sand dune ridges, possible storm-surge barriers, portable storm barriers, and a host of other schemes, including buyouts of threatened homes. The Bloomberg plan was a precursor of the "Rebuild by Design"competition described below.

Several plans for the city's response to rising seas were submitted to the Rebuild by Design competition, four of which were accepted, and one was funded. The four winners were:

1   The "BIG U." This will include the construction of a 10-foot-high berm named the 10-mile "BIG U" all around lower Manhattan,

from 57th Street on the west side to 42nd Street on the east side. Its price tag is estimated at $3 billion (federal, state, and local money), and a 2-mile start on the BIG U, extending south from East 25th Street, is underway.

2 Waterproof Lower Manhattan. This would be done by raising buildings, putting generators on higher floors, waterproofing utilities, and rebuilding garages, streets, and parks to hold water during storm events. Buildings not raised could abandon or fortify the first two floors and simply accept tidal and storm surge flooding. This method also makes sense for other cities, such as Boston.

3 The Barrier Island. This idea was to create a 40-mile-long chain of barrier islands ten miles offshore that would break the waves before they reached the shore. It was a ludicrous idea because a constant huge supply of sand would be required to maintain islands located in such an unstable location: in "deep water" and with regard to current and future sea levels. It's hard to understand how this was considered a winning design.

4 The "Living Breakwaters." This would involve construction of a 4,000-foot-long chain of breakwaters along portions of Manhattan, probably made of rock, designed to reduce storm wave energy. They are called "living breakwaters" because salt marsh and oysters would be planted behind the breakwaters. For the time being, at least, the BIG U wins.

One spectacular proposal that was not accepted was to build a five-mile-long storm-surge barrier (something like the Ike Dike in Galveston) that would extend from the peninsula at Sandy Hook, New Jersey, to the Rockaways, New York. There would be at least three floodgates to hold back storm surges and, at the proper time, release the floods from the Hudson River. One problem with this method is that success depends on the river flood and the storm surge occurring at different times. If that doesn't happen, the floods will be greater than ever. The proposal has not been approved and probably won't be in the future. The high cost of building the barrier would present another problem for the city.

Clearly any effective response to sea level rise will cost a huge amount of money in New York and the rest of America. New York is a wealthy city and may be able to afford these solutions, but a strong possibility exists that the cost of protecting some other cities from sea level rise may well be prohibitive. New York has the advantage of having large areas within the city at high elevation and is built largely on solid, hard-rock ground (except for the filled margins) as opposed to New Orleans, which resides on the muddy Mississippi Delta.

Boston

Downtown Boston is built on fill, land pumped or trucked to the city margins to build out into Boston Harbor (totaling 5,250 acres). (Other American cities with substantial fill areas are New York, Newark, and San Francisco.) This fill makes up about one-sixth of the city area in Boston and is the largest such urban fill land in the United States. Today, it appears that the ocean is coming back to take over the places where the harbor was filled. All told, five waterfront neighborhoods are particularly threatened with sea level rise, including much of downtown Boston.

The Deer Island Wastewater Treatment Plant, which is at low elevation in Boston Harbor, serves more than two million people in 43 communities in the Boston area. The loss of this facility in a storm or by sea level rise will clearly be a disaster.

Forecasts indicate that $80 billion worth of real estate and 90,000 residents in Boston are threatened by sea level rise "in coming decades," or so say the proponents of a giant sea barrier costing tens of billions of dollars that would extend the four miles between Hull and Deer Island. The wall would rise at least 20 feet above low tide and would have gates to allow shipping in and out of the harbor. The goal of the wall and gates would be to hold back high tides and storm surges. The probable inspiration for this, the Texas Ike Dike, as well as the rejected New York storm-surge barrier, is the Rotterdam Maeslantkering storm-surge barrier in the Netherlands, on another sinking delta like New Orleans. The Dutch barrier is much larger. It extends 66 feet above normal water levels and is one of the largest (and most costly) movable steel structures in the world.

The flooding of the subways of New York in Hurricane Sandy brought

FIGURE 9.5 The Deer Island Wastewater Treatment Plant, which is at low elevation in Boston Harbor, serves more than 2 million people in 43 communities in the Boston area. The loss of this facility in a storm or by sea level rise will clearly be a disaster. *Photo credit: Paul VanDerWerf, Flikr, Creative Commons 2.0.*

sharply home the possibility of a future storm aided by sea level rise flooding Boston's Big Dig, the road network underneath the city and harbor. According to the *Wall Street Journal*, this possibility became even more urgent when a 2016 winter storm pushed the water level in Boston Harbor to the highest ever recorded. About $196 million would be needed to storm-proof the Big Dig.

As noted earlier, this extreme endpoint of proposed Boston solutions to the rising seas is an approach also being considered for New York City and Galveston/Houston, among other cities—extreme in cost and extreme in attempting to combat sea level rise instead of working with the water as the Dutch try to do. Less extreme proposed solutions include building berms or levees around neighborhoods expected to flood, constructing drainage canals, designing parks and parking garages as areas to be flooded, and requiring buildings to be prepared for flooding—for instance, moving generators to upper floors. Last but not least, moving to safety—the retreat option—would in many cases be a good idea.

To Sum It Up

It is clear that cities on the East and Gulf coasts are far more endangered from sea level rise than those on the West Coast. At least this is true within this century. It is also clear that there is a great deal of variation around the United States in the degree of preparation underway to reduce the impact of expected flooding and intensification of storms. Many cities make the claim that they are working to reduce the impact of global climate change, but often that refers to emission reduction of greenhouse gasses. A number of coastal towns claim that they are preparing for sea level rise by enforcing building elevation and setback regulations, enforcing building codes, and providing timely warnings to the community. All these things are fine, but they do not directly address the sea level rise, other than to protect buildings in a short-term sense. Preparation should include design of engineering methods (e.g., seawalls, levees, raising or moving of buildings, nourishment) and (most important) preparation for the possibility of retreating or abandoning some sections of a city. Retreating from some low-lying areas can allow for water retention there during flood events.

In general, it can be said that governments of American cities are not taking sea level rise seriously enough. Among the eight cities that we briefly discuss, New York, Charleston, and Norfolk seem to be the furthest along in planning and are even beginning to fund and carry out initial engineering schemes. New York found its way, thanks to Hurricane Sandy. Miami, the most endangered major city, seems to have been finally awakened by severe nuisance flooding. Yet, large, upscale construction projects continue unabated in that city. Authorities in Tampa/St. Petersburg, another extremely at-risk city, show few signs of real efforts beyond report writing. It's too soon to tell, but Houston citizens probably have learned that they must take immediate steps to make the city safer from future flooding, having learned that allowing totally unrestricted development (the most atrocious example was allowing construction of houses in flood reservoirs) produced a disaster in Hurricane Harvey. Almost certainly, Houston, Tampa/St. Petersburg, Miami, New Orleans, Charleston, Norfolk, the San Francisco Bay Area, and parts of Los Angeles do not have happy long-term futures.

UNDER WATER

· NATIONAL FLOOD INSURANCE
AND CLIMATE GENTRIFICATION ·

The National Flood Insurance Program (NFIP) started in the 1960s with the best intentions. It seemed like a good idea at the time, but with fifty years of hindsight, we see unanticipated consequences that hound the financial sustainability of the program and make the response to sea level rise even more challenging.

Flood insurance is required for all properties within FEMA-designated Special Flood Hazard Areas (known as zones A and V) that are purchased with federally backed mortgages. Insurance is required for properties in these two zones, known as the "100-year" flood zone, where there is a 1 percent or higher chance of flooding (though the frequency of flooding in most areas belies this 100-year event designation). The NFIP offers lower premiums through its Community Rating System program to communities that adopt mitigation measures such as changes in building codes intended to reduce flood damage. However, few communities participate in this program because the incentives to do so are so small, reports Margaret Peloso in her 2017 book *Adapting to Rising Sea Levels.*

The NFIP was created to fill an insurance void left when private insurers stopped providing flood insurance due to a series of catastrophic floods in the early twentieth century. After Hurricane Betsy devastated the Louisiana coast in 1965, Congress eventually instituted the program in 1968. The NFIP's debt has ballooned over the past several decades because the premiums do not accurately reflect the risks. In a 2008 report, the Government Accountability Office blamed this failure on FEMA's reliance on outdated flood probability data. FEMA's flood maps were based on historic flood data and did not take into account future changes that may impact flooding, such as increased intensity of storms and coastal erosion. The Biggert-Waters Flood Insurance Reform Act of 2012 directed FEMA to consider projections of sea level rise and erosion when updating their flood maps.

Regularly updating flood maps is essential during this age of climate change. The increasing obsolescence of the FEMA flood maps has become clear in the face of record-intensity hurricanes and rain events. With the record-breaking rainfall from Hurricane Harvey, Houston had suffered three 500-year floods within the space of three years. It was a clear sign that the FEMA flood maps were essentially useless, since they assumed that property above the 100-year storm elevation was low risk. Houston's flood problems, in part, come from a change in the nature of the ground surface within the city. The recent big floods reflected the urbanized shift from vegetation and soil to impermeable concrete and asphalt—a change that means less of the rainwater is absorbed by the soil, ensuring worse flooding. According to a University of California at Davis study of satellite images, fully half of the flooded areas in the Houston vicinity were outside the FEMA flood designations (both the "100-year" and "500-year" flood zones).

Current estimates place the NFIP's debt at between $25 and $35 billion. In 2017, President Trump signed a disaster relief bill providing $16 billion in debt relief to the NFIP. So, how did a program designed to provide affordable (subsidized) insurance for structures in flood-prone areas in exchange for flood mitigation go awry? For one thing, the NFIP encouraged greater development in hazardous areas because it offers subsidized rates that minimize the financial risk to homeowners. Few would take the chance of building an expensive home alongside the beach if the NFIP wasn't there to cover the repair and rebuilding costs. When the NFIP was created in the 1960s, many coastal homes were relatively modest. Now, much of the U.S. coast is dominated by dense development, often with multistory, multi-bedroom second homes. So a combination of increased population and increasingly costly structures has resulted in a significant increase in the land value.

A *Yale Environment 360* report reveals that the value of land and structures along the New Jersey coast rose from less than $1 billion in 1960 to more than $170 billion by 2017. The availability of federal flood insurance encourages building in hazardous areas that may not have been developed in the absence of insurance. Further, insurance encourages rebuilding in those same areas and often in the same exact manner after a house is damaged in a flood event rather than providing funds for relocation to less hazardous areas or requiring adaptation measures such as raising houses on pilings when doing postflood repairs.

One of the inherent flaws in the NFIP is that insurance is required only in the high-risk flood zones, known as the Special Flood Hazard Areas. By failing to require flood insurance in areas outside of the high-risk zones, the NFIP created a high-risk insurance pool. By their nature, floods tend to involve entire communities or at least neighborhoods, so when flooding occurs there is likely widespread damage. If insurance covers only those areas most likely to be hit by the floods, it does not spread the risk among lower-risk flood areas. So, the premiums taken in will likely be inadequate to cover the damages from flooding. Before the late 2017 tax bill eliminated the individual healthcare mandate, the Affordable Care Act sought to avoid a similar problem by requiring people to buy medical insurance or face a penalty. That way there would be a balance of the healthy and the sick paying into insurance programs. For real estate, because only the structures at highest risk are required to get flood insurance, the NFIP does not have a significant enough number of less-at-risk insurers paying premiums to balance out the risk. Many of the areas flooded during Hurricane Harvey were outside the FEMA-designated high-risk flood areas. Residents in those areas should also be required to have flood insurance to protect against catastrophic losses because there is a nonnegligible risk of flood.

Another problem for the program is the fact that many policyholders let their insurance lapse. The Union of Concerned Scientists' report "Overwhelming Risk: Rethinking Flood Insurance in a World of Rising

Seas" indicates that only an estimated 15 to 25 percent of at-risk properties in the Northeast Special Flood Hazard Areas were insured when Hurricane Sandy roared ashore in 2012. Nationally, an estimated 18 percent of homes in flood-zone areas currently are insured. Banks are failing to enforce the requirement of flood insurance once the mortgage is obtained or if it is resold to other banks. The problem of uninsured homeowners will likely get much worse as the owners face annual increases in premiums.

The Barker and Addicks Reservoirs

Jeremy Boutor, a resident of Houston, Texas, had to abandon his house during Hurricane Harvey, carrying his 10-year-old son on his back and wading through the floodwaters. Before he was forced to flee after more than a foot of water entered his home, Boutor, like many others in Houston, was nervously watching the television news coverage of the hurricane and the flooding. Boutor was shocked to learn from the television that the home he had been renting for the past two years was actually located inside the Barker reservoir. It turns out Boutor wasn't alone. Many of the residents whose homes were flooded inside the Barker and Addicks reservoirs were unaware their residences lay inside an area designed to flood and retain floodwaters. However, reporter Neena Satija and her colleagues knew that development had been allowed inside the two reservoirs. Satija was investigating the story in the year before Hurricane Harvey and details her investigation in the *Texas Tribune*, as well as on the blog/podcast *Reveal*, for the Center for Investigative Reporting.

Following disastrous flooding in downtown Houston in 1935, the U.S. Army Corps of Engineers built two reservoirs to retain water during flood events. Unfortunately, the Corps purchased only 24,500 acres when it knew 8,000 more acres could flood within the reservoirs. At that time, the unpurchased 8,000+ acres were mostly agricultural, rice, and hay fields. The Corps' failure to purchase all the land inside the reservoirs over half a century ago set up the 2017 disaster that occurred during Harvey. In addition, allowing development within the reservoirs impaired the reservoirs' ability to absorb water by introducing paved surfaces. Some 14,000 homes lie within the two reservoirs, and during Harvey 5,138 of those homes flooded. Fears over the integrity of the aging dams, which were long recognized as hazardous, resulted in the decision to release waters to avoid catastrophic dam collapse. This release flooded more homes outside of the reservoirs. Lawsuits have been filed by homeowners, and the blame should fall on multiple parties, including the Army Corps of Engineers for failing to purchase all the land that could be flooded, as well as on the city of Houston and Harris County for allowing unchecked urban sprawl, in general, and construction of homes within reservoirs, in particular. Satija reveals that some homes within the reservoir that flooded during Harvey have subsequently sold. It seems that some Houston residents share the mistaken belief with many coastal residents who purchase property in hazardous areas that it "can't happen to me," or they possess the "development is 'Too Big to Flood' mentality" discussed in chapter 9. Also discussed in chapter 9 is the saga of engineer Charles Glen Crocker who in 1992 warned officialdom that subdivisions were being built in a future lake and was ignored.

FEMA bases the 100-year storm zone on past storm history. But things change, as in the case of Houston's expanding, poorly draining concrete and asphalt. That's not the only thing that changes. Global climate change is expected to produce larger and possibly more frequent storms with increased precipitation, which will also negate the flood maps. The fundamental mistake in FEMA's maps is that they are based on storm history, a probability approach that is the typical basis for insurance ratings. This backward-looking approach is becoming increasingly insufficient. One thing climate change robs you of is the ability to easily predict outcomes based on the type of events that have occurred in recent decades. Potentially, the changes implemented by the Biggert-Waters Act may help FEMA's mapping become more accurate.

The Biggert-Waters Act was a long-overdue attempt to reform the NFIP. One of its most ambitious measures was an increase in the NFIP insurance rates to more accurately reflect actual risk. The act sought to cure the NFIP's significant debt issues by allowing premiums to rise to sound actuarial levels. However, when homeowners caught wind of their increased rates, the political blowback was so intense that the act was significantly diluted. The Homeowner Flood Insurance Affordability Act of 2014 repealed flood insurance rate increases that would have significantly raised premiums. Peloso points out that some New Jersey homeowners in areas prone to flooding and damage from waves found out their insurance premiums would increase from $900 per year to over $30,000 per year. These shocking numbers reveal just how heavily subsidized the NFIP premiums are. It begs the question: how many would build in these risky areas if they had to pay insurance rates that accurately reflected the risks of building in coastal flood zones?

The Homeowner Flood Insurance Affordability Act still calls for premium rate increases, but introduces them more slowly by capping most annual increases at 18 percent. Insurance rates for nonprimary residential properties, business properties, Severe Repetitive Loss properties, and substantially damaged/substantially improved properties face annual increases of 25 percent until they reach full-risk rates.

Another major problem that has become clear over the 50-year existence of the NFIP is the high cost of Severe Repetitive Loss properties. Repetitive-loss properties are defined as properties that have four or more claims exceeding $5,000 each or two or more claims where the total value of the claims exceeds the fair market value of the property. A 2017 report from the Natural Resources Defense Council reveals that the NFIP insures over 30,000 Severe Repetitive Loss properties that have flooded an average of five times, on average every two to three years. These structures represent just 0.6 percent of the 5.1 million properties insured by the NFIP, but they account for a whopping 9.6 percent of all damages paid between 1978 and 2015, a total of $5.5 billion. A 2010 GAO report found that repetitive loss properties account for only 1 percent of the NFIP policies but an astounding 5 to 30 percent of all claims.

The numbers may vary by source, but it is clear that these repetitive-loss properties are a major financial drain on the NFIP. In all, there are 11,000 NFIP-insured properties where the cumulative payouts far exceed the value of the structures. The roll call of such events reveals an insurance insanity that would never be accepted by a private insurer. Some of the neighborhoods flooded in Houston in 2017 by Hurricane Harvey had been flooded several times since 2010. As pointed out by the *Washington Post*, one Houston home, valued at $72,000, has received over $1 million in payouts. A home near Baton Rouge, Louisiana, valued at $55,921, has flooded an astounding 40 times, resulting in $428,379 in claims. The *New*

York Times reported that a Spring, Texas, home valued at $42,000 has been repaired at least 19 times to the tune of $912,732.

Another significant problem for these beleaguered homeowners is that the repeat flooding will make it difficult if not impossible to sell these homes. Because of the owner's or builder's initial bad decision in choice of building site, they are trapped in hazardous locations—unable to sell their homes even as the value of the homes decreases with each recurring flood. "If I had a choice, I would sell," said the 65-year-old Leni-Anne Shuchter, of Pequannock, New Jersey, who dreams of retiring to Arizona or Nevada. "I don't need to deal with this anymore. [But] the reality of selling is nil." The *Washington Post* reported in July 2016 that Ms. Shuchter's property had flooded numerous times and that once she had to be rescued from a neighbor's roof during a 1984 flood. She planned to use a FEMA grant to elevate her home. Increasingly, when it comes to multiple insurance payments and beach nourishment tax increases, inland dwellers argue that they didn't build in a very unsafe place so they should not have to take the tax hit to support those who did. Most of us were wiser in our choice of home site.

The Homeowner Flood Insurance Affordability Act repealed flood insurance rate increases that applied to properties that were uninsured as of July 6, 2012, and to properties purchased after that date. The act also allowed purchasers of a property with a current NFIP policy to assume the policy at existing rates, referred to as "grandfathering." Grandfathering locks in lower premiums that do not reflect the actual flood risk. Biggert-Waters sought to eliminate the practice of grandfathering, including the elimination of allowing homeowners to keep their already existing premiums even if updated flood maps call for increased rates.

The Biggert-Waters Act ended the extension of rate subsidies to owners of repetitive-loss properties who refuse mitigation assistance following a major disaster. Mitigation assistance often comes in the form of the Stafford Act, which provides federal money to help rebuild communities after presidential-declared emergencies. The Stafford Act allows for contributions to state and local governments for repair of public facilities, direct assistance for repair and rebuilding of private housing, and funding for the repair and replacement of federal facilities. The Stafford Act does encourage some mitigation measures. For instance, the federal share of repair and rebuilding costs to local governments, which is typically 75 percent, can be reduced if a public facility was damaged in the past ten years by the same type of event and the owner failed to take mitigation measures. Also, with regard to homeowners, those who receive assistance under the Stafford Act to replace a damaged structure must maintain flood insurance coverage. In addition, the property owners who are required to maintain coverage after receiving assistance have a duty to inform buyers that the requirement to maintain coverage is assigned to the buyer. One critique of the Stafford Act is that it emphasizes quick recovery following a disaster and allows rebuilding in high-hazard areas in basically the same manner as had occurred before the storm. Basically, the act works against the need to retreat from hazardous locations that will become more hazardous as sea level rises.

NFIP Reform Ideas

In their 2013 report "Overwhelming Risk: Rethinking Flood Insurance in a World of Rising Seas," the Union of Concerned Scientists (UCS) suggests several reforms for the NFIP. First of all, they call for the NFIP premiums to actually reflect the risks to coastal properties and for an income-based voucher or rebate program to assist lower-income homeowners to afford the higher insurance rates. The UCS also suggests reducing payments for repetitive-loss properties to discourage building in floodplains and the cycle of rebuilding these houses over and over after floods. They also would like removal of "grandfathering" provisions that allow some property owners to pay even lower (more subsidized) premiums. This would help the premiums reflect the actual risk. Also, the UCS suggests increased home buyouts, an idea we agree with because purchasing damaged homes and leveling them ensures that no more public funds will be squandered insuring homes in hazardous areas. And the UCS wants to require that flood maps be shared with buyers to make sure they are actually aware of the flood risks. The group wants changes in regulations that would mandate flood insurance in high-risk areas and encourage the purchase of flood insurance in the areas that may not be highest risk but are still at risk of flooding. Too many residents are allowed to let their insurance lapse after purchasing homes, which adds to the financial problems of the NFIP.

It remains to be seen whether Congress has the stomach to implement further flood insurance reform. Reform is sorely needed, and a primary focus should be tailoring it to mitigate future damage from floods and to discourage construction in hazardous areas while encouraging removal of buildings in flood-prone areas. What is clear is that the NFIP and the Stafford Act have failed to adequately prevent construction in hazardous areas. Both have been factors in overdevelopment of high-risk coastal areas.

While the Stafford Act allows federal funding for disaster assistance, typically, federal funds are made available in a sharing agreement, where state governments provide 25 percent of the costs to the federal share of 75 percent of the pie. Former FEMA director Craig Fugate and his Trump-appointed successor, Brock Long, have both backed the idea of a "disaster deductible." In this approach, states would have to chip in a certain amount of money, a deductible, before federal monies would be made available. FEMA could reduce the deductible for those states taking actions to mitigate disasters, such as implementing stronger building codes or prohibiting development in hazardous areas. Many of FEMA's regulations are designed to encourage mitigation, but the fact is that these regulations are inadequate. Perhaps Fugate's disaster deductible could be a useful tool in helping states prepare for sea level rise.

Climate Gentrification

Sea level rise is perhaps the most predictable outcome of climate change. Whether you believe global warming is caused or significantly influenced by human activity, or you buy the argument that the huge science support for global climate change is misplaced and the climate change we are experiencing is part of some natural cycle, the fact of the matter is that the sea is rising and shall continue to do so. Time and tide wait for no human, no matter what said human believes with regard to the origin of climate change.

One predictable outcome from sea level rise is climate gentrification. As the seas rise, the real estate mantra of "Location! Location! Location!"

will shift to "Elevation! Elevation! Elevation!" And with that shift, some of the people located on higher grounds, particularly low-income folks, many of whom rent, will be shifted right out of their neighborhoods by increased housing prices. As defined by activist Valencia Gunder, climate gentrification is when low-income and marginalized communities are displaced by factors related to climate change and sea level rise.

The term *climate gentrification* is attributed to Jesse Keenan, a lawyer and lecturer with Harvard University's Graduate School of Design. Keenan sees evidence in survey data that middle-income people are leaving places such as Miami Beach and other places with nuisance flooding that makes it difficult to get around at high tides. Where might these folks be heading? The low-income neighborhoods of Little Haiti and Liberty City, famously featured in the film *Moonlight*, are both situated at 10 feet above sea level, which in Miami is the high ground. That's where you will see climate gentrification in Miami. Anthropologist Hugh Gladwin has been predicting for years that this will occur, though he told *The Root* that it is too early to tell for certain that it is climate-based gentrification that is beginning to drive out the poorer residents at higher elevation.

Regardless, we feel the gentrification of higher-elevated low-income neighborhoods is merely a hurricane away. Had Hurricane Irma swept up the east side of Florida instead of the west side, the gentrification would likely be in full swing. Writing in the *Atlantic*, Matt Vasilogambros argues that developers are already quietly buying up land on higher ground. Vasilogambros writes about the Little Farm community, a 15-acre trailer park a few blocks from Little Haiti, which was bought by a Chinese company that evicted the residents in order to develop the land. Many of the 100 residents owned their trailers, but their homes were so old the trailers were no longer mobile. The residents were offered between $1,500 and $2,500 to vacate their homes voluntarily, and some received an $8,000 payout after a lawsuit alleging violation of the Florida Mobile Home Act, which calls for relocation studies to be performed ahead of closure agreements. Trailer owners rented space for as low as $500 per month and will be hard-pressed to find another place to live at such a low cost in South Florida.

The rich low-lying residents will increasingly move to the higher-elevation poorer neighborhoods as they grow frustrated with dealing with storm fatigue, the hassles of sunny day flooding, and a future full of increased flood insurance costs. They will buy up rental houses, displacing low-income renters, and, as demand increases on the dwindling numbers of rentals, rents will increase and further displace the poor. The developers won't be far behind as they recognize that low-cost rentals can be reconfigured or rebuilt to serve wealthier clientele.

One cruel irony is that people of color in South Florida were historically kept from desirable waterfront communities, such as Miami Beach, due to the old Jim Crow laws. Per a *Miami New Times* article, in the early to mid-twentieth century, the only blacks allowed in Miami Beach were those employed to serve the wealthy whites. A 1936 ordinance required seasonal workers at hotels, many of them black, to register with the police and to carry IDs. In 1952, Miami Beach police arrested seventeen African American bus riders for failing to comply with the law. With rising waters making life along the water more difficult, marginalized communities occupying the high ground will be pushed out by rising real-estate demands and costs, and we will lose the urban culture that makes these Miami neighborhoods unique.

In 2018, citing research by Keenan, the city of Miami decided to study the issue of climate gentrification and to look into methods to stabilize tax rates in order to allow residents who wish to remain in their neighborhoods to do so. The city has dedicated $4 million from the Miami Forever Bond to assist at-risk residents in fixing up their homes. *The Miami Herald* identified the following communities as low-income, high-elevation areas at risk: Liberty City, Overtown, Allapattah, Little Havana, and Little Haiti. It appears that Miami is the first city in the U.S. to study the topic of climate gentrification.

In a 2017 article published online by Cory Schouten, Jesse Keenan lays out three pathways to climate gentrification. Keenan's prescient pathways are as follows:

1   Low-risk properties surge in value, fueling a migration from high-exposure areas, causing displacement. This isn't just a sea-level issue: California's wildfires, for instance, are likely to lead to significant changes in how real estate is valued. "Anything related to climate change," Keenan explained. "Low exposure (to climate-related risks) is the determinant."

2 Living in high-exposure areas gets so expensive (think taxes, insurance) that only rich people can live there, pushing historically mixed-income areas (such as Miami Beach, Florida, and Hampton Roads, Virginia) to become more exclusive.

3 Government investments in resilience have the unintended consequence of boosting land and property values that wind up displacing populations. Sea-level fortifications on the Lower East Side of Manhattan could have this effect, Keenan said.

We would add postdisaster gentrification to Keenan's list. This particular form of gentrification is made possible by destruction of homes by hurricanes and storms and has been going on for decades. Rapid increases in property values postdestruction have happened in New Orleans on a grand scale, and thousands of lower-income residents, many of them African American, fled in the aftermath of the storm, never to return. The website FiveThirtyEight.com publicizes an estimate that more than 175,000 black residents left the city after Katrina in 2005, and more than 75,000 never returned. Some of this exodus was the result of policy decisions. For instance, the federal Road Home rebuilding program based payments on the appraised value of damaged properties (which was often lower in black neighborhoods) rather than on the cost of repairing the properties. This left many black families without enough money to rebuild.

Also, when the Louisiana state legislature took over the New Orleans schools, they fired all 4,600 teachers, along with hundreds of other staffers. While some were rehired, thousands of mostly black school employees lost their jobs as the state moved to initiate public charter schools. A 2017 study from the Education Research Alliance at Tulane University shows that half of the teachers fired never worked in Louisiana public schools again. The percentage of black public school teachers dropped from 71 percent of the entire faculty in 2003 to under 50 percent by 2014.

Many of New Orleans's low-lying neighborhoods are predominantly black. Some residents were initially dislodged by the floodwaters, but others were finally forced out by shortsighted postdisaster policies. Now, New Orleans is smaller, whiter, and more expensive, and in danger of losing the very culture that draws so many people to the city in the first place.

When a storm demolishes a seaside house, that empty lot is sometimes sold for more money, and the vanished home is replaced by a more expensive home or condominium. The authors saw this phenomenon play out in Waveland, Mississippi, where the home of the authors' parents and grandparents was destroyed by Hurricane Katrina in 2005. One year post-Katrina, small seaside lots, some with concrete slabs where the former houses had stood, were going for $800,000 (as advertised on signs posted on nearby live oaks that survived the storm). Storms effectively clear the way for urban renewal, and the homes that replace those lost tend to be grander and more expensive and, in many cases, multi-occupant dwellings, creating greater population density in hazardous waterfront areas. This destroy-and-rebuild cycle ensures future natural disasters.

In a 2018 article for Bloomberg, Christopher Flavelle, author of many excellent pieces on climate-change issues, also suggests a pattern of storm-damaged properties being replaced by more expensive structures and pushing out low-income residents. Flavelle points out that Florida Keys building codes prohibit replacing or substantially repairing damaged mobile homes, leaving residents of nearly 1,000 trailers and RVs destroyed by 2017's Hurricane Irma with the options of finding sturdier, but more expensive lodging, replacing or repairing their homes and hoping the code officials don't notice, or leaving the Keys. Flavelle also cites U.S. Department of Housing and Urban Development (HUD) data showing that with no federal requirement to replace or repair public housing units that receive money from HUD, many units damaged or destroyed by storms are gone for good. Texas did not replace all of the 1,260 public housing units destroyed by Hurricanes Ike and Dolly in 2008. Galveston has replaced only about half of the 569 units lost to the 2008 storms. Hurricane Harvey (2017) damaged or destroyed close to 1,500 public housing units in the Houston area alone.

The wake of 2017's busy hurricane season reveals an ugly side to storm recovery and how it leads to gentrification in the Caribbean, including the U.S. Virgin Islands. In a piece posted to a Society of Ethnobiology web blog on October 19, 2017, approximately one month after Hurricane Irma wreaked havoc in the Caribbean, the authors note that stateside Virgin Islanders working as real estate agents had already received calls asking whether cheap land was available.

Jesse Keenan published a 2018 study in the journal *Environmental Research Letters* in which he found that the prices of single-family homes in lower elevations of Miami-Dade County have been appreciating at a slower pace than homes at higher elevations since approximately 2000. Quoted in the *Miami Herald*, Keenan states that "it really stood out that low elevation properties essentially serve now as inferior investments." Reviewing 500,000 house and condominium sales, Keenan found that homes vulnerable to sea level rise sold for an average price 7 percent lower than homes that are less at risk to sea level rise. The disparity for homes that would be inundated with a one-foot sea level rise was even greater. Those homes sold for 18 percent less than homes at higher elevation.

Asaf Bernstein and colleagues also researched coastal properties, using prices from the Zillow Transaction and Assessment Dataset along with the National Oceanic and Atmospheric Administration's sea-level-rise calculator to determine which properties would be inundated by a six-foot rise in seas. Their 2018 study also found that homes exposed to sea level rise sell for approximately 7 percent less than equivalent homes in areas that are not threatened by the rising seas.

Scientists affiliated with the First Street Foundation published a 2018 peer-reviewed study in the journal *Population Research and Policy Review* tallying the accrued losses in real estate due to the threat of sea level rise and the reality of increased nuisance flooding in the Miami-Dade area. The authors found that properties projected to be inundated with tidal flooding in 2032 have lost $3.08 each year on each square foot of living area, and properties near roads that will be inundated with tidal flooding in 2032 have lost $3.71 each year on each square foot of living area. Between 2005 and 2016, the Miami-Dade area suffered losses in real estate value totaling an estimated $465 million. Lest one think the property value losses are limited to South Florida, First Street Foundation expanded the analysis to cover all of Florida, Georgia, South Carolina, North Carolina, and Virginia by examining over 5.5 million real estate transactions in these states and extrapolating the results to 12.2 million properties. They concluded that there has been a total home value loss of $7.4 billion since 2005.

Florida leads the way with $5.42 billion in lost property values, followed by South Carolina with losses of $1.11 billion, North Carolina with

TABLE 10.1 Property Value Loss

State	# Properties with lost value	Amount of value lost
Georgia	11,360	$15 million
Virginia	47,035	$280 million
North Carolina	81,908	$582 million
South Carolina	92,775	$1.1 billion
Florida	384,548	$5.42 billion
Total properties that lost vValue	616,626	
Total property value loss	$7.4 billion	

Property value loss due to sea level rise from 2005 to 2017.

SOURCE: Adapted from Press Release, First Street Foundation, July 25, 2018. "As the seas have been rising, home values have been sinking." FirstStreet.org.

losses of $582 million, Virginia with losses of $280 million, and Georgia with $15 million in lost value. So, the waters are already steadily lapping away at the value of coastal homes. Given that land will give way to water and that properties across the entire U.S. coast will be inundated, it is inevitable that the real estate bubble will burst. Those who reside along the coast should ask themselves, can they afford to take the financial hit that will come when their home loses money as the oceans continue to rise?

WHAT YOU CAN DO
ABOUT SEA LEVEL RISE

It is difficult but necessary to come to grips with the ultimate outcome of living with Nature at the shoreline.

The Truths of the Rising Sea

- If you can see the sea, the sea can see you.
- The sea will continue to rise for centuries to come. It won't stop by 2100. Consider it a 400-year problem.
- The current scientific consensus expects about a three-foot rise by 2100.
- The current scientific consensus high-end projection is for a six-foot rise by 2100.
- On coastal plains, a small sea level rise will produce inundation far inland.
- On mountainous coasts, the amount of shoreline retreat will be variable but usually much less than on coastal plain shores.
- Sea level rise, nuisance flooding, and storm surges are interrelated.

- Building homes or businesses and any large buildings in risk zones is a disfavor to the next generations.
- The shoreline cannot be held still in the long run. Once you start holding the shoreline still, you can't stop trying.
- Holding the shoreline still increases development.
- Holding the shoreline still causes great environmental harm.
- Seawalls on eroding shorelines will destroy the beach.
- Beach nourishment is a short-term erosion solution.
- FEMA flood maps are often of limited use for risk evaluation in an age of climate change.
- Millions will move from the U.S. coasts by 2100.
- "Safe" neighborhoods may nonetheless be trapped by low-lying, flooded access and escape roads. Destruction of water and wastewater plants will force an immediate retreat from "safe" towns.

How to Live with the Coast

The following discussion addresses the questions of what an individual moving to the coast or one who already is there should know and should do in a time of rising sea level. The discussion centers on home sites close to the shoreline—close enough that the occupants of a house should, while they live there, experience directly or indirectly the impact of sea level rise, tidal flooding, storm surges, and shoreline erosion. All of these are interrelated phenomena. For example, as the seas rise, storm surges, tidal flooding, and shoreline erosion increase. As shorelines retreat as a result of sea level rise, storm surges penetrate farther inland, and so it goes. The impact of sea level rise in the twenty-first century could extend a few hundred yards on the West Coast along some hard-rock shorelines and a hundred miles or more along some U.S. East and Gulf shores.

The City View

- There's a Dutchman in your future. The Netherlands is the world's leader in dealing with water and has recently shifted from a view of keeping waters out by any means to the idea of

FIGURE 11.1 A traffic jam on the evacuation route as residents try to evacuate on I-45 north of Houston, Texas, before Hurricane Rita in 2005. Clearly all lanes of the interstate should have been opened for storm evacuation traffic. Notice the fender bender in the lower left portion of the photo. A similar situation existed in Charleston, South Carolina, where only half of an interstate was used for citizens fleeing Hurricane Floyd (which never came ashore). The traffic jam was so extreme there that many escapees spent the night sleeping in their cars. *Photo by Ashram, Creative Commons—I-45 and Louetta [CC BY 2.0 (https://creativecommons .org/licenses/by/2.0)], via Wikimedia Commons.*

living with the water—accommodating floodwaters on some lands while using mammoth infrastructure to protect dense urban areas. This scheme will inevitably come to a shoreline near you.

- Coastal cities must choose which areas can be defended and which areas can be used in a way that can help to accommodate or temporarily hold floodwaters during flood events.

- Cities must also review their vulnerability. At what elevation are the wastewater treatment plants, reservoirs, road maintenance facilities, energy infrastructure, hazardous material storage, schools, police stations, jails, and access and escape roads? What will happen if the wastewater and drinking water plants are out of commission, and what can be done about it? Identify the infrastructure at risk and work now to move it or protect it from floodwaters.

- There will be winners and losers, and we should take steps to protect those citizens most vulnerable. Much like the fact that poorer nations often bear the brunt of climate change and sea level rise, poor people will lose out in the scramble for higher ground unless cities specifically work to avoid displacement of the poor.

- Cities should move rapidly toward renewable energies. Not only will this help reduce greenhouse gas productions that are causing climate change, but this will enable quicker rebounding from natural disasters such as hurricanes. If you want resilient cities, make the buildings within the cities energy independent.

- In those areas that will experience flood events down the road, plan now to require flood-friendly first floors. Consider placing parking on the lower levels of buildings to minimize damage to residences and businesses from floods. Consider elevated entrances to first floors of buildings to discourage waters from coming inside.

- Cities must implement and enforce building standards designed to help accommodate future floods. Elevation can stave off the effects of floods, buying time, but ultimately it is necessary to identify which areas to protect and which areas to cede to the waters.

- Retreat, in some degree, will be in the future of all coastal cities, whether it is small scale (New York) or, in some places, large scale (Miami), depending on elevation, geological factors, and the continued rise of seas with the passage of centuries, rather than decades.
- It is important to recognize that while we tend to plan for the relatively short term, sea level rise is not going to stop at the year 2100. Especially if we don't reduce greenhouse gas emissions, sea level rise will continue, and it will be absolutely catastrophic.
- What does catastrophic sea level rise mean for coastal cities? Eventually, it will mean abandonment for some cities. Florida, in particular, faces an existential threat from sea level rise. New Orleans and much of the wetlands-laden Louisiana coast may also wash away eventually.
- The refugees are coming. States may wish to craft entirely new urban centers to which the sea-level-rise-displaced coastal city dwellers may relocate. Newly crafted, carefully planned cities could provide a chance to avoid the pitfalls of haphazardly planned development.

Moving to the Coast

Questions You Should Ask

- Is it really important to you to move to a site where sea level rise, nuisance flooding, storm surges, and shoreline retreat will become part of your life? Even if your site is a "safe" one, you may pay high taxes bailing out those who were less prudent in site choice.
- What happened in the last storm? What happened in the "big one"? Look up storm history on the internet. What is the maximum storm surge height predicted for your area by the National Oceanic and Atmospheric Agency (NOAA)?
- How successful was community evacuation in the last storm?
- What is the erosion rate of the shoreline or the bluff?

- What is the local expected or assumed sea level rise 50 years from now? Seventy-five years from now? Consult online tools to see maps with the expected inundation from various levels of sea level rise. Don't forget that as sea levels rise, so shall the nuisance flooding levels.
- Is your town or state cognizant of sea level rise and the expected future intensification of storms as the oceans warm and rise? Are they doing something about it?
- Does the town enforce setback lines (if they have them) and building codes for high wind construction? Do they encourage or forbid seawalls?
- Is the beach a principal reason for your move to the coast? What is the likelihood of its preservation into the future? What are the community's plans in this regard?

Personal Considerations

- How many years do you expect to live in the house? Can you afford to lose money when the house is destroyed or when the coastal real-estate-market bubble bursts in the face of sea level rise?
- Do you hope that your children will inherit the house you plan to purchase? If so, is this house in a location where it is likely to survive long enough?
- Are you retired and your children don't want the house/condo? Stay right there and enjoy it, but make sure you're prepared to evacuate quickly.
- Looking ahead to future tidal flooding around your neighborhood and on surrounding access roads, how will that affect your lifestyle and access to schools, shopping, church, and evacuation?
- How will rising seas and tidal flooding affect the monetary value of your house or condo?
- Can you afford federal and private insurance? The cost of federal flood insurance will undoubtedly rise.
- Do you have the money needed to cope with future tidal flooding? Can you start saving money for this purpose?

Choosing Your Home Site

- If you can see the sea, the sea can seize you.
- Don't depend solely on the FEMA risk maps for your choice of a home site. Look at storm history. Talk to the neighbors and visit city hall.
- The common Realtor's pronouncement concerning factors that most affect property values is "location, location, location"— but if you're moving to the coast, the three most important factors are "elevation, elevation, elevation."
- Don't buy beachfront property unless you're willing to pay a lot for insurance, beach nourishment, seawall, and bulldozing costs, and can afford to lose the house in a storm.
- The "safe" elevation for a home varies from location to location. The safest will be above the elevation of the highest expected storm surge. In Bay St. Louis, Mississippi, the potential storm surge is more than 30 feet. In Nags Head, North Carolina, 10 feet is a high storm surge (see NOAA info).
- Are the access roads safe? Will they flood well before the town does, thus cutting off escape?
- Where is the wastewater plant? Will it flood before the town does, forcing immediate abandonment?
- Make sure your house was or is built to code (to withstand high winds and floods). Some state codes are weak. Other states don't enforce them, and in some, codes don't exist (Texas, Alabama, and Mississippi). The books listed in Appendix C, the *Living with the Shore Series*, will help you understand building codes and what to look for.
- Add five feet to the storm surge expectation to be even safer, to account for the additional surge height caused by sea level rise.
- To increase the elevation of your home, build on stilts and don't give in to the temptation to add walls to the stilts to add ground-level rooms. Stilts are designed to allow water to flow underneath a home, so if you add walls it will obstruct the water and cause more damage to the home. Don't forget that

elevation doesn't address the high-wind problem—in fact, it makes it worse.

- Consider buying a second lot, higher and safer than your primary lot. The second lot could be the place to move your house—in case.
- If you plan to live on an island, how easy is it to evacuate? Is there just one bridge or one road for thousands of people?
- Make plans to evacuate your pets with you. Don't risk your life to stay with your pets during a storm and don't leave them behind.
- If you plan to move to an island community with beachfront high-rises, almost certainly you are facing a future with sea-walls on both the front side and the back side of the island, which means eventually there will be no recreational beach.

Appendix C is a list of the *Living with the Shore* books published by Duke University Press. These state-by-state books discuss the relative risks of various U.S. shorelines.

If You Are Already at the Coast

- If you are an alert homeowner, you should be able to answer most of the questions in the earlier "Questions You Should Ask" section. Take a look at them.
- Pay close attention to spring tide timing and particularly to schedules of king tides and the impacts from them that you can expect in your area. Consult websites for this information.
- If you recognize the problems you might have in the future with sea level rise, tidal flooding, storms, and shoreline retreat, you can do the following:
  - Raise your house.
  - Move house to a safer location.
  - Move your house to a higher elevation site.
  - Move to a new and safer condominium.
  - Sell your house and move to another town.
  - Retreat altogether from the region and start anew.

Maybe you should sell your house now while you can get a good price for it (before the real estate market tanks) and move inland or to a different town. Moving from Miami, Ft. Lauderdale, Tampa-St. Petersburg, New Orleans, the Mississippi Delta, Norfolk, and the Inner Banks right now might make sense. Moving out of South Florida now will beat the rush of climate refugees that will begin to move in a few decades. It will be better to move in a planned fashion rather than to flee in the aftermath of a catastrophe.

Storm or flood survivors consistently are most disappointed if they have lost family photos and records. Put them in a box that can easily be taken with you—or scan them into the computer and put them on a flash drive you can easily carry, or upload to the cloud.

Know your evacuation route. Experience has shown that older people have a more difficult time in evacuations. Remember, though you may be at a high elevation, the surrounding land may be low, and you might be trapped by a storm surge or even by nuisance flooding. And don't wait to be ordered to evacuate. If there's even a small chance that a hurricane coming up the coast will hit your location, the time has come to visit "Aunt Susie in Topeka." Create a "bug out" kit of supplies you may need in an evacuation. Don't forget your medications and eyeglasses. Flee early in response to sea level rise rather than with the fleeing crowd.

Start saving money to cope with future flooding.

Escaping Nature

The November 8 California Camp Fire that destroyed almost the entire town of Paradise (population 26,000), brought into focus, once again, the problem of escaping natural disasters. Although a fast-moving forest fire is different than a flood or storm surge, the two types of disasters share the critical problem of escape of inhabitants who have the misfortune of being in the way. In the town of Paradise, escape routes had already been identified and publicized because of a previous smaller forest fire. But when the big Camp Fire came, evacuation was impossible on three of the potential routes due to heavy smoke.

Along the Gulf and Atlantic lower coastal plains are many small towns located on "topographic highs" which offer protection from floods. But often the access roads to the towns are at lower elevations which flood in storms, blocking escape—just like the smoke in California.

Another critical problem facing escapees in Paradise was overcrowding of remaining escape roads, significantly slowing down evacuation. The same has happened frequently to people escaping hurricanes. For example, roads were jammed with escapees from hurricanes Floyd (1999), Rita (2005; figure 11.1), Katrina (2005), and Irma (2017). Inevitably adding to the problem, cars ran out of gas and road-blocking fender benders occurred. In all of these storms, two out of four lanes of an interstate were kept free for emergency vehicles, a bad decision that led to the two useable lanes being too congested for movement. In forest fires, there must be a lane open for firefighters, but no such critical need exists in most hurricanes. The emergency vehicles can come in after the storm. There is not much they can do in the midst of high winds and flying debris. In the case of Hurricane Floyd, many escapees from Charleston, South Carolina, on the two open interstate lanes didn't escape and spent the night sleeping in their cars. As it turned out, Floyd didn't come ashore in Charleston as predicted. The traffic jam in Rita was chiefly caused by "shadow evacuees," individuals who evacuated in spite of no need or requirement to do so.

In cities, vertical evacuation to upper floors of well-built high-rise buildings should be emphasized where feasible. There is also the Bangladesh solution, which is akin to the vertical solution. In that country, which is so vulnerable to cyclone floods, the government has constructed a number of shelters to which local residents can retire in a storm.

Our Super-Cautious Advice to Our Family Members

*To a Relative Living in a Beachfront High-Rise Condo
on a Barrier Island on Any Coast*

In the long run, the chances are that on the beach side and the back side, there will be a high, continuous seawall without a recreational beach. So instead of enjoying the beach, you will enjoy the sea air. Roads to and from the barrier island will probably be frequently blocked by tidal floods and permanently blocked by sea level rise. Best to live above the second floor. Eventually you will have trouble getting to the doctor, church, school, or work, and maintenance people won't be able to get to work to help you.

To a Relative Living in Seattle, Washington

From Highlands to Shilshole Bay and Magnolia Bluff, the bluffs are often quite unstable and will become more so as sea level rises. Don't believe the consulting engineers' assurances of safety that may be required for your bank loans. Look for yourself and learn the history of local bluff failures. Please move back from the bluff-top to a safer location so you won't have to fret whenever a heavy rain or high winds and high waves occur. There are lots of high-elevation home sites here, but they don't necessarily have a view of water. Supposedly the Ballard Locks will hold back the rising sea and storm surge on lakes Union and Washington in the long run of several decades. A six-foot sea level rise probably will top the locks. Check with city hall.

To a Relative Living on the Beach in Malibu, California

Malibu Colony is one of the many beaches that make up the 21 miles of shoreline in Malibu. This development has received some publicity because it consists of a line of costly buildings, some occupied by famous people, right on the beach and fronted by seawalls. There is no question that when the big one comes, many of these buildings will be damaged or destroyed, and this is just a matter of time. As the sea rises, the hazard for this community virtually on the beach will increase immeasurably. There is also a hazard here from mudslides. You could be caught between the devil in the mudslides and the deep blue sea that is rising to get you. We would advise you to sell (while the selling is good) and leave the area for higher ground.

To a Relative Living in Bay St. Louis/Waveland, Mississippi

Please move back immediately to a new site more than 30 feet in elevation. Don't trust the elevation on the city maps. The storm surge potential here is 30 feet—it happened in Hurricane Katrina (2005) and almost happened in Hurricane Camille (1969). Some houses on the beachfront have been destroyed twice but have been replaced. To live on the beachfront here is madness. (We have a cousin who lives here— away from the beach.)

To a Relative Living on the Western Half of Dauphin Island, Alabama

Please move immediately. This place is unsafe.

To a Relative Living in South Florida

If you live in Ft. Lauderdale, Marco Island, Tampa/St. Petersburg, any other coastal Florida town, or the Florida Keys, please think about moving north in the next few years. No hurry; take your time, consider the present-day value of your house, remember that the value of your property will soon drop, and consider where you might move and work in a different location. If you live in Miami or Miami Beach, you should move a bit more rapidly. Remember the study (chapter 9) that shows that of the 25 U.S. cities most vulnerable to coastal flooding, all but five are in Florida.

To a Relative Living in Myrtle Beach, South Carolina

Myrtle Beach is not a barrier island; instead, it is a stretch of mainland. The community is atop an ancient barrier island (formed 120,000 years ago) that is as high as 60 feet in elevation. Get off the beachfront, and if you can find a home site, live on the back side of Myrtle Beach with the "high" elevations. Eventually, as sea level rises, the roads leading to the high portions of Myrtle Beach will be flooded. Best to give up the sea view and move inland, educating yourself before you choose a home site. North Myrtle Beach and Cherry Grove, both north of Myrtle Beach proper, are low in elevation and are likely to be overwashed in big storms. Behind Cherry Grove are numerous low-elevation finger canals, heavily developed, with a strong potential pollution problem, especially in warm summer months (so not a particularly good place to be).

To a Relative Living in Nags Head, North Carolina

As you already must know, Nags Head has a lot of storms. It has some of the biggest waves on the East Coast of the United States and has a serious shoreline erosion problem. On top of that, most of Nags Head is very low, with the exception of some areas like the giant sand dunes and Nags Head Woods. Evacuation could be difficult because to get off the island to a reasonably high area on the mainland, you must go for miles through the Inner Banks, which are low and susceptible to flooding in storms. If you continue to live here, don't live on the beachfront and do be prepared to evacuate well ahead of any storm.

To a Relative Living in Plymouth, North Carolina, in the Inner Banks

You are living in one of many small towns in an area of particular danger from sea level rise even though you are miles from the sea. You are, however, close to Albemarle Sound. The problem is the low elevation— already tidal flooding is temporarily isolating some towns. Please consider moving inland to a higher elevation out of the Inner Banks. In a decade or two, you will be forced to move, so why not beat the crowd before the value of your house tanks?

To a Relative Living in Oxford, Maryland, on the Chesapeake Bay

You are lucky to be living in a town that is very aware of the sea-level-rise hazard and is planning to take measures to reduce flooding. They have even increased taxes for this purpose. But keep in mind that Oxford, like many other Chesapeake Bay communities, is at a low elevation. Since you are at retirement age, perhaps you can hang out for a while, but keep an eye on tidal flooding, and remember a hurricane moving up the bay could cause a lot of problems. Be prepared to evacuate along your preplanned route, and keep an eye open for a possible move to a safer site. (We have a cousin who lives here.)

To a Relative Living at Seaside Heights, New Jersey

This is part of a very low and narrow island with high-density development. Streets run completely perpendicular to the shoreline, making perfect overwash passes during storms. The closely packed houses can be destroyed or damaged by storms and will provide debris that can cause damage to adjacent buildings. The dune protection currently in place is very ephemeral and will be more so as sea level rises. We recommend that you move to a less dense and higher-elevation area away from the shorefront, or even better, move to the mainland.

To a Relative Living on Rockaway Beach, New York

Rockaway Beach is at the southwestern tip of Long Island. The Rockaways are among the most threatened parts of New York City. One foot of sea level rise will cause significant tidal flooding that gradually will become total inundation. With 3 feet of sea level rise, the beach will be gone and may well be replaced by a large seawall. With 6 feet, the entire Rockaway Peninsula will be under water. You'll be paying taxes to protect the city. All in all, we recommend you move away from Rockaway. Take your time and choose a new home site carefully, after reading the extensive literature available on sea level rise in New York.

A Final Thought

Like it or not, we will retreat from most of the world's nonurban shorelines in the not-very-distant future. Cities are another matter. Some cities, like Miami, will be abandoned, and millions of refugees will flee South Florida. Other cities, like New York, have some elevations well above the anticipated sea level rise, and we may only be forced to retreat from the city margins. Boston has low elevations but perhaps can be protected by a massive seawall and other measures. The U.S. Pacific Coast is highly variable in its geologic framework, ranging from sand spits to high rocky cliffs, so the need for retreat will vary considerably.

Retreat options can be characterized as either difficult or catastrophic. We can plan now and retreat in a strategic and calculated fashion, or we can worry about it later and retreat in tactical disarray in response to devastating storms. In other words, we can walk away methodically, or we can flee in panic.

· APPENDIX A ·

Global Delta Population Displacement Potential by 2050

EXTREME: Greater than 1 Million People

Nile, Egypt
Ganges-Brahmaputra, Bangladesh/India
Mekong, Vietnam

HIGH: 1 Million to 500,000 People

Mississippi, United States
Godavari, India
Yangtse, China

MEDIUM: 500,000 to 5,000 People

Grigalva, Mexico	Shatt Al Arab, Iraq
Oronoco, Venezuela	Indus, Pakistan
Sao Francisco, Brazil	Krishna, India
Rhine, Netherlands	Mahanadi, India
Sabou, Morocco	Pearl, China
Moulouya, Morocco	Chao Phraya, Thailand
Volta, Ghana	Mahakam, Indonesia
Niger, Nigeria	Senegal, Senegal/Mauritania

SOURCE: *The Intergovernmental Panel on Climate Change (IPCC)*

*The Economic and Environmental Price of Holding
the Shoreline Still with Hard Stabilization*

SEAWALLS	**Shore-parallel walls**
Economic price	$2,000 to $10,000/ft.
Environmental price	Extreme—eventual complete loss of beach

GROINS	**Shore-perpendicular walls**
Economic price	$250 to $6,500/ft.
Environmental price	Updrift beach widened; downdrift beaches lost

Costs of groins vary depending on materials used

Stone	$1,200–$6,500/ft.
Concrete, sheet steel	$4,000–$5,000/ft.
Timber	$3,000–$4,000/ft.
Geotextille (large sand bag)	$250–$1,000/ft.

JETTIES	**Shore-perpendicular structures next to inlet**
Economic price	Up to $16,500/ft
Environmental price	Updrift beach widened; downdrift, extreme beach loss

BREAKWATERS	**Shore-parallel but offshore**
Economic price	Up to $10,000/ft
Environmental price	Trap sand as they widen local beaches but cause severe downdrift erosion

SOURCE: Dollar figures are from the U.S. National Park Service's 2016 Coastal Strategies Handbook.

Living with the Shore Book Series

State-Specific Shoreline Risk Books

The following is a list of books from the *Living with the Shore* series published by Duke University Press, written by experts from the various states and edited by Orrin Pilkey and William Neal. These state-specific books provide risk maps based on storms and shoreline erosion rather than sea level rise. Each book has maps showing the entire shoreline of the state and the risk to development, from extreme to moderate. The maps in these books, some of which are out of date, should still be useful for prospective homeowners. The general rule to follow is: don't live on the beachfront, no matter what the classification is in these books or on the FEMA maps.

- Bush, David M., Richard M. T. Webb, José G. Liboy, Lisbeth Hyman, and William J. Neal. 1995. *Living with the Puerto Rico Shore.*
- Bush, David M., Norma J. Longo, William J. Neal, Luciana S. Esteves, Orrin H. Pilkey, Deborah F. Pilkey, and Craig A. Webb. 2001. *Living on the Edge of the Gulf: The West Florida and Alabama Coast.*
- Bush, David M., William J. Neal, Norma J. Longo, Kenyon C. Lindeman, Deborah F. Pilkey, Luciana S. Esteves, John D. Congleton, and Orrin H. Pilkey. 2004. *Living with Florida's Atlantic Beaches: Coastal Hazards from Amelia Island to Key West.*
- Canis, Wayne F., William J. Neal, Orrin H. Pilkey Jr., and Orrin H. Pilkey Sr. 1985. *Living with the Alabama-Mississippi Shore.*
- Carter, Charles H., William J. Neal, William S. Haras, and Orrin H. Pilkey Jr. 1987. *Living with the Lake Erie Shore.*
- Clayton, Tonya D., Lewis A. Taylor Jr., William J. Cleary, Paul E. Hosier, Peter H. F. Graber, William J. Neal, and Orrin H. Pilkey Sr. 1992. *Living with the Georgia Shore.*

- Doyle, Larry J., Dinesh C. Sharma, Albert C. Hine, Orrin H. Pilkey Jr., William J. Neal, Orrin H. Pilkey Sr., David Martin, and Daniel F. Belknap. 1984. *Living with the West Florida Shore.*
- Griggs, Gary, and Lauret Savoy, eds. 1985. *Living with the California Coast.*
- Kelley, Joseph T., Alice R. Kelley, Orrin H. Pilkey Sr., and Albert A. Clark. 1984. *Living with the Louisiana Shore.*
- Kelley, Joseph T., Alice R. Kelley, and Orrin H. Pilkey Sr. 1989. *Living with the Coast of Maine.*
- Komar, Paul D. 1997. The Pacific Northwest Coast: *Living with the Shores of Oregon and Washington.*
- Lennon, Gered, William J. Neal, David M. Bush, Orrin H. Pilkey, Matthew Stutz, and Jane Bullock. 1996. *Living with the South Carolina Coast.*
- Mason, Owen, William J. Neal, Orrin H. Pilkey, Jane Bullock, Ted Fathauer, Deborah Pilkey, and Douglas Swanston. 1997. *Living with the Coast of Alaska.*
- McCormick, Larry R., Orrin H. Pilkey Jr., William J. Neal, and Orrin H. Pilkey Sr. 1984. *Living with Long Island's South Shore.*
- Morton, Robert A., Orrin H. Pilkey Jr., Orrin H. Pilkey Sr., and William J. Neal. 1983. *Living with the Texas Shore.*
- Neal, William J., W. Carlyle Blakeney Jr., Orrin H. Pilkey Jr., and Orrin H. Pilkey Sr. 1984. *Living with the South Carolina Shore.*
- Nordstrom, Karl F., Paul A. Garès, Norbert P. Psuty, Orrin H. Pilkey Jr., William J. Neal, and Orrin H. Pilkey, Sr. 1986. *Living with the New Jersey Shore.*
- Patton, Peter C., and James M. Kent. 1992. *A Moveable Shore: The Fate of the Connecticut Coast.*
- Pilkey, Orrin H., William J. Neal, Stanley R. Riggs, Craig A. Webb, David M. Bush, Deborah F. Pilkey, Jane Bullock, and Brian A. Cowan. 1998. *The North Carolina Shore and Its Barrier Islands.*
- Pilkey, Orrin H. Jr., Dinesh C. Sharma, Harold R. Wanless, Larry J. Doyle, Orrin H. Pilkey Sr., William J. Neal, and Barbara L. Gruver. 1984. *Living with the East Florida Shore.*
- Terich, Thomas A. 1987. *Living with the Shore of Puget Sound and the Georgia Strait.*
- Ward, Larry G., Peter S. Rosen, William J. Neal, Orrin H. Pilkey Jr., Orrin H. Pilkey Sr., Gary L. Anderson, and Stephen J. Howie. 1989. *Living with Chesapeake Bay and Virginia's Ocean Shores.*

· REFERENCES ·

Prologue

Dawson, Alastair. 2018. *Introducing Sea Level Change*. Edinburgh: Dunedin Academic Press Ltd.

Gertner, Jon. 2018. "The Race to Understand Antarctica's Most Terrifying Glacier." *Wired*. December 10. Accessed December 11, 2018. https://www.wired.com /story/antarctica-thwaites-glacier-breaking-point/.

Intergovernmental Panel on Climate Change. 2018. "Global Warming of 1.5°C: An IPCC Special Report on the Impacts of Global Warming of 1.5°C above Pre-industrial Levels and Related Global Greenhouse Gas Emission Pathways, in the Context of Strengthening the Global Response to the Threat of Climate Change, Sustainable Development, and Efforts to Eradicate Poverty." Summary for Policymakers. This Summary for Policymakers was formally approved at the First Joint Session of Working Groups I, II and III of the IPCC and accepted by the 48th Session of the IPCC, Incheon, Republic of Korea, October 6. Accessed October 8, 2018. http://www.ipcc.ch/report/sr15/; http://report.ipcc.ch /sr15/pdf/sr15_spm_final.pdf.

McCullough, David. 1992. *Brave Companions: Portraits in History*. New York: Simon and Schuster Paperbacks.

NOAA. U.S. Department of Commerce. 1975. "The Coastline of the United States." 1975 revision. Brochure. Accessed April 5, 2018. https://shoreline.noaa.gov/_pdf /Coastline_of_the_US_1975.pdf.

Pilkey, Orrin H., and Robert Young. 2009. *The Rising Sea*. Washington, DC: Island Press.

Plumer, Brad, and Lisa Friedman. 2018. "Trump Team Pushes Fossil Fuels at Climate Talks. Protests Erupt, but Allies Emerge, Too." *New York Times*, December 10. Accessed December 11, 2018. https://www.nytimes.com/2018/12/10/climate /katowice-climate-talks-cop24.html.

SeaLevelRise.org. 2018. "The Future of Sea Level Rise: Sea Level Rise Is Speeding Up." SeaLevelRise.org. Accessed December 21, 2018. https://sealevelrise.org /forecast/.

Sekich-Quinn, Stefanie. 2018. "2018 State of the Beach Report Card Released."
Surfrider Foundation. December 13. Accessed December 21, 2018. https://
www.surfrider.org/coastal-blog/entry/2018-state-of-the-beach-report-card
-released.

Showstack, Randy. 2018. "Arctic Undergoing Most Unprecedented Transition in
Human History." *Eos 99*, December 12. Accessed December 15, 2018. https://doi
.org/10.1029/2018EO111981.

USGCRP. 2017. *Climate Science Special Report: Fourth National Climate Assess-
ment, Volume I*, edited by D. J. Wuebbles, D. W. Fahey, K. A. Hibbard, D. J.
Dokken, B. C. Stewart, and T. K. Maycock. Washington, DC: U.S. Global
Change Research Program.

USGCRP. 2018. *Impacts, Risks, and Adaptation in the United States: Fourth National Climate Assessment, Volume II*, edited by D. R. Reidmiller, C. W. Avery,
D. R. Easterling, K. E. Kunkel, K. L. M. Lewis, T. K. Maycock, and B. C. Stewart.
Washington, DC: U.S. Global Change Research Program. https://doi.org
/10.7930/NCA4.2018.

Chapter 1. Flee the Sea

Anderson, Charles. 2017. "New Zealand Considers Creating Climate Change
Refugee Visas." *Guardian*, October 31. Accessed November 8, 2017. https://
www.theguardian.com/world/2017/oct/31/new-zealand-considers-creating
-climate-change-refugee-visas.

Bahr, Len. 2009. "The (Real) Trouble with Harry!" People Issues, *LaCoastPost:
All Things Coastal*, June 29. Accessed October 8, 2017. http://lacoastpost.com
/blog/?p=9321.

Blum, Michael D., and Harry Roberts. 2009. "Drowning of the Mississippi Delta
Due to Insufficient Sediment Supply and Global Sea-Level Rise." June. *Nature
Geoscience* 2 (7): 488–91. https://doi.org/10.1038/ngeo553.

Cornell University. 2017. "Rising Seas Could Result in 2 Billion Refugees by 2100."
ScienceDaily, June 26. Accessed July 7, 2017. https://www.sciencedaily.com
/releases/2017/06/170626105746.htm.

Cronin, William B. 2005. *The Disappearing Islands of the Chesapeake*. Baltimore:
Johns Hopkins University Press.

Dagenais, Travis. 2017. "Rising Seas, Distressed Communities, and 'Climate
Gentrification': Jesse M. Keenan Talks Miami in *Vice, Scientific American*."
Cambridge, MA: Harvard Graduate School of Design. August 14. Accessed
January 27, 2018. http://www.gsd.harvard.edu/2017/08/rising-seas-distressed
-communities-and-climate-gentrification-jesse-m-keenan-talks-miami-in-vice
-scientific-american/.

Environmental Justice Foundation (EJF). 2017. "Beyond Borders: Our Changing
Climate—Its Role in Conflict and Displacement" (Climate Campaign: https://

ejfoundation.org/what-we-do/climate), November. Accessed January 12, 2018. https://ejfoundation.org/resources/downloads/BeyondBorders.pdf.

Environmental Justice Foundation (EJF). 2017. "Protecting Climate Refugees." November. Accessed January 12, 2018. https://ejfoundation.org/what-we-do/climate/protecting-climate-refugees.

Flavelle, Christopher. 2017. "Louisiana, Sinking Fast, Prepares to Empty Out Its Coastal Plain." Bloomberg *Politics*, December 22. Accessed December 30, 2017. https://www.bloomberg.com/news/articles/2017–12–22/louisiana-sinking-fast-prepares-to-empty-out-its-coastal-plain.

Ghose, Tia. 2016. "American Counties at Risk of Flooding from Climate Change." *Live Science*, March 14. Accessed June 20, 2017. https://www.livescience.com/54040-cities-at-risk-climate-change.html.

Hauer, Mathew E. 2017. "Migration Induced by Sea-Level Rise Could Reshape the U.S. Population Landscape." *Nature Climate Change* 7 (May): 321–27. Published online April 17, 2017. Accessed May 30, 2017. https://doi.org/10.1038/nclimate3271. http://www.nature.com/nclimate/journal/v7/n5/full/nclimate3271.html?foxtrotcallback=true.

Hauer, Mathew E., Jason M. Evans, and Deepak R. Mishra. 2016. "Millions Projected to Be at Risk from Sea Level Rise in the Continental United States." *Nature Climate Change* 6:691–95. Published online March 14, 2016. Corrected online March 21, 2016. Corrected online April 22, 2016. Accessed June 20, 2017. https://doi.org/10.1038/nclimate2961.

Hine, Albert C., Don P. Chambers, Tonya D. Clayton, Mark R. Hafen, and Gary T. Mitchum. 2016. *Sea Level Rise in Florida: Science, Impacts, and Options*. Gainesville: University Press of Florida.

Krol, Debra Utacia. 2018. "In Louisiana, a Plan to Relocate the Country's First 'Climate Refugees' Hits a Roadblock." *Environment, HuffPost*, March 23. Accessed March 26, 2018. https://www.huffingtonpost.com/entry/louisiana-climate-refugees-plan-roadblock_us_5ab402ade4b008c9e5f55c1b.

LA SAFE (Louisiana's Strategic Adaptations for Future Environments). 2017. "Changing with the Coast: Helping Louisiana Residents to Strategically Adapt, Plan and Explore Options in the Face of Coastal Erosion." Baton Rouge. Accessed October 20, 2017. https://lasafe.la.gov/.

McDaniel, Melissa, Erin Sprout, Diane Boudreau, and Andrew Turgeon. 2017. "Climate Refugee." *National Geographic*. Last updated June 17, 2011. Accessed June 25, 2017. https://www.nationalgeographic.org/encyclopedia/climate-refugee/.

McDougall, Dan. 2009. "The World's First Environmental Refugees." *Ecologist*, January 30. First appeared in the *Ecologist*, December 2007. Accessed October 22, 2017. http://www.theecologist.org/investigations/climate_change/269582/the_worlds_first_environmental_refugees.html.

Miami Herald, Sun Sentinel, Palm Beach Post, and WLRN Editorial Staffs. 2018.
"The Invading Sea: Can South Florida Be Saved?" *Miami Herald, Sun Sentinel,
Palm Beach Post*, and WLRN *News*, June 10. Accessed June 15, 2018. https://www
.theinvadingsea.com/contact-us/.

National Oceanic and Atmospheric Administration (NOAA). 2014. "'Nuisance
Flooding' an Increasing Problem as Coastal Sea Levels Rise," July 28. Updated
October 31, 2014. Accessed June 15, 2017. http://www.noaanews.noaa.gov/stories
2014/20140728_nuisanceflooding.html.

National Oceanic and Atmospheric Administration (NOAA). 2014. "Sea Level Rise
and Nuisance Flood Frequency Changes around the United States." June. Silver
Spring, MD: NOAA Technical Report NOS CO-OPS 073.

Northern Arizona University. 2017. "Gulf Coast: Biloxi-Chitimacha-Choctaw In-
dians; Rising Tides." Accessed November 3, 2017. http://www7.nau.edu/itep
/main/tcc/Tribes/gc_choctaw.

Northern Arizona University. 2017. "Tribal Profiles." Accessed November 3, 2017.
http://www7.nau.edu/itep/main/tcc/Tribes.

Pilkey, Orrin H., Linda Pilkey-Jarvis, and Keith C. Pilkey. 2016. *Retreat from a
Rising Sea: Hard Choices in an Age of Climate Change*. New York: Columbia
University Press.

Riggs, Stanley R., Dorothea V. Ames, Stephen J. Culver, and David J. Mallinson.
2011. *The Battle for North Carolina's Coasts: Evolutionary History, Present Crisis,
and Vision for the Future*. Chapel Hill: University of North Carolina Press.

Rush, Elizabeth. 2017. "Harvey and Irma Are the New Normal. It's Time to Move
Away from the Coasts." *Washington Post Outlook*, September 15. Accessed
October 15, 2017. https://www.washingtonpost.com/outlook/irma-and-harvey
-are-the-new-normal-its-time-to-move-away-from-the-coasts/2017/09/15
/4ff2a61e-9971–11e7–87fc-c3f7ee4035c9_story.html?utm_term=.480a9a758b50.

Snider, Annie. 2017. "It's Not Going to Be All Right." *Politico Magazine*, September
1. Accessed October 10, 2017. http://www.politico.com/magazine/story
/2017/09/01/harvey-texas-louisiana-floods-relocation-215565.

Sun Sentinel Editorial Board. 2018. "Sea-Level Rise: The Defining Issue of the
Century—Editorial." *Sun Sentinel*, May 4. Accessed June 3, 2018. http://www
.sun-sentinel.com/opinion/editorials/fl-op-editorial-sea-level-rise-attention
-needed-20180503-story.html.

Taylor, Matthew. 2017. "Climate Change 'Will Create World's Biggest Refugee
Crisis.'" *Guardian*, November 2. Accessed November 7, 2017. https://www
.theguardian.com/environment/2017/nov/02/climate-change-will-create-worlds
-biggest-refugee-crisis.

Union of Concerned Scientists. 2017. "When Rising Seas Hit Home: Hard Choices
Ahead for Hundreds of Coastal U.S. Communities." Accessed November 2,
2017. http://www.ucsusa.org/RisingSeasHitHome.

U.S. Census Bureau. 2018. "Nevada and Idaho Are the Nation's Fastest-Growing States." Newsroom: Press Release, December 19. Accessed December 24, 2018. https://www.census.gov/newsroom/press-releases/2018/estimates-national-state .html.

Wall, Kim, Coleen Jose, and Jan Hendrik Hinzel. 2017. "On Standby: When You Leave the Marshall Islands, You Buy a One-Way Ticket." *Mashable.com.* Accessed March 1, 2018. https://mashable.com/2018/02/25/marshall-islands -climate-refugees/#TLC8xgo8TOq9.

Chapter 2. The End of the Inupiat Way of Life

Alaska Department of Fish and Game. 2018. "Community Subsistence Informa- tion System: CSIS." Juneau, Alaska. Accessed February 18, 2018. http://www .adfg.alaska.gov/sb/CSIS/.

Baillargeon, Zoe. 2017. "On This Chilean Island the Whole Community Helps Move Your House." *Atlas Obscura.* Accessed November 8, 2017. https://www .atlasobscura.com/articles/moving-houses-of-chiloe.

Benen, Steve. 2017. "Widespread Corruption Allegations Add to Trump World's Troubles." *Washington Post*, October 5. Accessed October 6, 2017. http://www .msnbc.com/rachel-maddow-show/widespread-corruption-allegations-add -trump-worlds-troubles?cid=eml_mra_20171005.

Bronen, Robin. 2013. "Climate-Induced Displacement of Alaska Native Communi- ties: Brookings-LSE Project on Internal Displacement." Alaskan Immigration Justice Project, January 30. Accessed November 5, 2017. https://www.brookings .edu/wp-content/uploads/2016/06/30-climate-alaska-bronen-paper.pdf.

Brubaker, Michael, James Berner, Jacob Bell, and John Warren. 2011. "Climate Change in Kivalina, Alaska: Strategies for Community Health." Alaska Native Tribal Health Consortium (ANTHC), January. Accessed October 6, 2017. http:// www.anthc.org/chs/ces/climate/climateandhealthreports.cfm.

Choi, Charles Q. 2011. "Inuit Dog Sleds Help Measure Arctic Sea Ice." NBCNews .com, January 18. Accessed January 3, 2018. http://www.nbcnews.com /id/41006342/ns/technology_and_science-science/t/inuit-dog-sleds-help -measure-arctic-sea-ice/#.WkvEpXlryM8.

Clement, Joel. 2017. "Read Joel Clement's Resignation Letter." *Washington Post*, October 4. Accessed January 23, 2018. https://apps.washingtonpost.com/g /documents/national/read-joel-clements-resignation-letter/2566/.

Cooke, Kieran. 2017. "Rising Warmth Risks Arctic Dogs' Survival." *Climate News Network*, August 22. Accessed January 3, 2018. https://climatenewsnetwork .net/22918-2/.

Cox, Meki. 1997. "Hawaiian Families Reclaim Their Native Home." *Hawai'i— Independent and Sovereign*, March 17. Accessed February 15, 2018. https://www .hawaii-nation.org/kahikinui.html.

Goode, Erica. 2016. "A Wrenching Choice for Alaska Towns in the Path of Climate Change." *New York Times*, November 29. Accessed October 5, 2017. https://www .nytimes.com/interactive/2016/11/29/science/alaska-global-warming.html ?mcubz=0.

Hawai'i Climate Change Mitigation and Adaptation Commission. 2017. *Hawai'i Sea Level Rise Vulnerability and Adaptation Report*. Prepared by Tetra Tech, Inc., and the State of Hawai'i Department of Land and Natural Resources, Office of Conservation and Coastal Lands, under the State of Hawai'i Department of Land and Natural Resources Contract No. 64064. Accessed January 18, 2018. https://climateadaptation.hawaii.gov/wp-content/uploads/2017/12/slr-Report _Dec2017.pdf.

Hayes, Tryck N., and URS Corporation. 2006. "Relocation Planning Project, Master Plan, Kivalina, Alaska." U.S. Army Corps of Engineers, Alaska District. Contract No./Order No. DACW85–03-D-0006–0003, Delivery Order No. 0007, June. Accessed March 15, 2016. http://www.poa.usace.army.mil/Portals/34/docs /civilworks/reports/KivalinaMasterPlanMainReportJune2006.pdf.

Hilburg, Jonathan. 2017. "Houston Unveils Post-Harvey Downtown Master Plan." *Architect's Newspaper*, November 13. Accessed January 26, 2018. https:// archpaper.com/2017/11/houston-harvey-master-plan/.

Hughes, Zachariah. 2018. "In Doomed Alaska Town, Hunters Turn to Drones and Caribou as Sea Ice Melts." *Guardian*, March 2. Accessed March 2, 2018. https://www.theguardian.com/environment/2018/mar/02/alaska-climate -change-indigenous-hunting.

Kennedy, Merrit. 2016. "Threatened by Rising Seas, Alaska Village Decides to Relocate." *The Two Way: National Public Radio*, August 18. Accessed November 5, 2016. http://www.npr.org/sections/thetwo-way/2016/08/18/490519540 /threatened-by-rising-seas-an-alaskan-village-decides-to-relocate.

Klouda, Naomi. 2017. "Denali Commission Directed to Work on Shutdown Plan." *Alaska Journal of Commerce*, April 5. Accessed August 15, 2017. http://www .alaskajournal.com/2017–04–05/denali-commission-directed-work-shutdown -plan.

Knoblauch, Jessica A. 2018. "Climate Change Forces Quinault Tribe to Seek Higher Ground." *EarthJustice*, March 12. Accessed August 3, 2018. https:// earthjustice.org/blog/2018-march/climate-change-forces-the-quinault-tribe -to-seek-higher-ground.

Marino, Elizabeth. 2011. "Case 2: Flood Waters, Politics, and Relocating Home: One Story of Shishmaref, Alaska." In *North by 2020: Perspectives on Alaska's Changing Social-Ecological Systems*, edited by A. Lovecraft and H. Eicken, 261–63. Fairbanks: University of Alaska Press.

Marino, Elizabeth. 2015. *Fierce Climate, Sacred Ground: An Ethnography of Climate Change in Shishmaref, Alaska*. Fairbanks: University of Alaska Press.

Osborne, Emily, Jackie Richter-Menge, and Martin Jeffries, eds. 2018. NOAA Arctic Report Card 2018. Accessed December 12, 2018. https://www.arctic.noaa.gov/Report-Card.

Pilkey, Orrin H., with original batiks by Mary Edna Fraser. 2003. *A Celebration of the World's Barrier Islands.* New York: Columbia University Press, 235–41.

Pilkey, Orrin H., Linda Pilkey-Jarvis, and Keith C. Pilkey. 2016. *Retreat from a Rising Sea: Hard Choices in an Age of Climate Change.* New York: Columbia University Press, 98–102.

Quinault Indian Nation Community Development and Planning Department. 2017. "Taholah Village Relocation Master Plan," June 26. Accessed August 3, 2018. http://www.quinaultindiannation.com/planning/FINAL_Taholah_Relocation_Plan.pdf.

Semuels, Alana. 2015. "The Village That Will Be Swept Away." *Atlantic.* Accessed June 5, 2017. https://www.theatlantic.com/business/archive/2015/08/alaska-village-climate-change/402604/.

U.S. Climate Resilience Toolkit. 2015. "Relocating Kivalina: Case Study." Adapted from the report "Climate Change in Kivalina, Alaska: Strategies for Community Health," published by the Alaska Native Tribal Health Consortium, and the blog post "Re-Locate Receives 2015 ArtPlace America Grant," published by Re-Locate Kivalina. Accessed December 12, 2017. https://toolkit.climate.gov/case-studies/relocating-kivalina.

U.S. Government Accountability Office. 2015. "Denali Commission: Options Exist to Address Management Challenges: Report to Congressional Requestors." GAO 15–72, March 25. Accessed December 12, 2017. https://www.gao.gov/assets/670/669225.pdf.

Chapter 3. Lord Willing and the Creek Don't Rise

Harrington, Samantha. 2017. "Georgia Island Confronts 'Blue Sky' Floods." *Yale Climate Connections*, April 21. Accessed December 26, 2017. https://www.yaleclimateconnections.org/2017/04/georgia-island-confronts-blue-sky-floods/.

Karegar, Makan A., Timothy H. Dixon, Rocco Malservisi, Jürgen Kusche, and Simon E. Engelhart. 2017. "Nuisance Flooding and Relative Sea-Level Rise: The Importance of Present-Day Land Motion." *Scientific Reports* 7, no. 11197 (September 11). Accessed February 18, 2018. https://doi.org/10.1038/s41598-017-11544-y. https://www.nature.com/articles/s41598-017-11544-y.

LeMonte, Joshua J., Jason W. Stuckey, Joshua Z. Sanchez, Ryan Tappero, Jörg Rinklebe, and Donald L. Sparks. 2017. "Sea Level Rise Induced Arsenic Release from Historically Contaminated Coastal Soils." *Environmental Science and Technology* 51, no. 11 (May 4): 5913–22. Accessed July 23, 2018. https://doi.org/10.1021/acs.est.6b06152.

Milman, Oliver. 2017. "Atlantic City and Miami Beach: Two Takes on Tackling

the Rising Waves." *Guardian*, March 20. Accessed March 22, 2017. https://www
.theguardian.com/us-news/2017/mar/20/atlantic-city-miami-beach-sea-level
-rise.

Moftakhari, Hamed R., Amir AghaKouchak, Brett F. Sanders, and Richard A.
Matthew. 2017. "Cumulative Hazard: The Case of Nuisance Flooding." *Earth's
Future* 5 (February 22): 214–23. Accessed November 8, 2017. https://doi.org
/10.1002/2016EF000494. https://agupubs.onlinelibrary.wiley.com/doi/abs
/10.1002/2016EF000494.

Pilkey, Orrin H., and Linda Pilkey-Jarvis. 2007. *Useless Arithmetic: Why Environmental Scientists Can't Predict the Future.* New York: Columbia University Press.

Samenow, Jason. 2018. "Federal Report: High-Tide Flooding Could Happen 'Every
Other Day' by Late This Century." *Washington Post*, March 28. Accessed June
19, 2018. https://www.washingtonpost.com/news/capital-weather-gang
/wp/2018/03/28/federal-report-high-tide-flooding-could-happen-every-other
-day-by-late-this-century/?noredirect=on&utm_term=.d515137ce54a.

Smialowski, Brendan. 2018. "High-Tide Flooding in the U.S. Is Twice What It Was
30 Years Ago, NOAA Says." CBS *News*, June 7. Accessed March 16, 2018. https:
//www.cbsnews.com/news/high-tide-flooding-in-the-u-s-is-twice-what
-it-was-30-years-ago-noaa/.

Staletovich, Jenny. 2016. "Miami Beach King Tides Flush Human Waste into Bay,
Study Finds." *Miami Herald*, May 16. Accessed July 22, 2017. http://www.miami
herald.com/news/local/environment/article77978817.html.

Sweet, William V., Greg Dusek, Jayantha Obeysekera, and John J. Marra. 2018.
*Patterns and Projections of High Tide Flooding along the U.S. Coastline Using a
Common Impact Threshold.* NOAA Technical Report NOS Co-Ops 086. Silver
Spring, MD: National Oceanic and Atmospheric Administration, U.S. Department of Commerce, National Ocean Service, Center for Operational Oceanographic Products and Services, February. Accessed March 16, 2018. https://
tidesandcurrents.noaa.gov/publications/techrpt86_PaP_of_HTFlooding.pdf.

Sweet, William V., Doug Marcy, Greg Dusek, John J. Marra, Matt Pendleton. 2018.
"2017 State of U.S. High Tide Flooding with a 2018 Outlook. Supplement to *State
of the Climate: National Overview for May 2018*," June 6. Accessed June 19, 2018.
https://www.ncdc.noaa.gov/monitoring-content/sotc/national/2018/may/2017
_State_of_US_High_Tide_Flooding.pdf.

Union of Concerned Scientists. 2017. "North Carolina Faces Chronic Inundation:
Fact Sheet." Accessed November 2, 2017. http://www.ucsusa.org/RisingSeas
HitHome.

Union of Concerned Scientists. 2018. "Underwater: Rising Seas, Chronic Floods,
and the Implications for U.S. Coastal Real Estate," June. Accessed June 24, 2018.
https://www.ucsusa.org/sites/default/files/attach/2018/06/underwater-analysis
-full-report.pdf.

UNC-TV Science. 2018. "Sea Level Rise Puts Pressure on Hyde County." *Peril and Promise: The Challenge of Climate Change. PBS*, February 5. Accessed February 6, 2018. http://www.pbs.org/wnet/peril-and-promise/2018/02/sea-level-rise-puts -pressure-on-hyde-county/.

U.S. Climate Resilience Toolkit. 2016. "U.S. Climate Resilience Toolkit," November 26. Accessed February 14, 2017. https://toolkit.climate.gov/topics/coastal-flood -risk/shallow-coastal-flooding-nuisance-flooding.

Viglucci, Andres, and Joey Flechas. 2017. "South Beach Wants to Save Art Deco Gems before the Seas Rise. One Solution: Jack Them Up." *Miami Herald*, April 14. Accessed June 18, 2017. http://www.miamiherald.com/news/local/community /miami-dade/miami-beach/article144575594.html.

Chapter 4. Dirty Waters and Worried Minds

Bacon, John. 2018. "Hurricane Maria Killed More Than 4,600 People—More Than 70 Times the Official Toll of 64, Study Says." *USA Today*, May 29. Accessed June 6, 2018. https://www.usatoday.com/story/news/nation/2018/05/29 /hurricane-maria-killed-thousands-puerto-rico/650942002/.

Bajak, Frank, and Lise Olsen. 2018. "Hurricane Harvey's Toxic Impact Deeper Than Public Told." Associated Press/*Houston Chronicle*. Chron.com, March 23. Accessed March 26, 2018. https://www.chron.com/news/texas/article /Hurricane-Harvey-s-toxic-impact-deeper-than-12772773.php.

Beachapedia. 2018. "Health Threats from Polluted Coastal Waters." Modified July 18. Accessed August 3, 2018. http://www.beachapedia.org/Health_Threats_from _Polluted_Coastal_Waters.

Brunkard, Joan, Gonza Namulanda, and Raoult Ratard. 2007. "Hurricane Katrina Deaths, Louisiana, 2005." *Disaster Medicine and Public Health Preparedness* 2, no. 4: 215–23. Published online April 8, 2013. Accessed February 27, 2018. https:// doi.org/10.1097/DMP.ob013e31818aaf55.

Bush, David M., Norma J. Longo, William J. Neal, Luciana S. Esteves, Orrin H. Pilkey, Deborah F. Pilkey, and Craig A. Webb. 2001. *Living on the Edge of the Gulf*. Durham, NC: Duke University Press.

Center for Investigative Reporting. 2018. "The Storm after the Storm." *Reveal* pod-cast, October 27. Accessed December 26, 2018. https://www.revealnews.org /episodes/the-storm-after-the-storm/.

Clayton, Susan, Christie Manning, Kirra Krygsman, and Meighen Speiser. 2017. "Mental Health and Our Changing Climate: Impacts, Implications, and Guid-ance." Washington, DC: American Psychological Association and ecoAmerica. Accessed November 22, 2017. http://www.apa.org/news/press/releases/2017/03 /mental-health-climate.pdf.

Dickerson, Caitlin. 2017. "After Hurricane, Signs of a Mental Health Crisis Haunt Puerto Rico." *New York Times*, November 13. Accessed November 23, 2017.

https://www.nytimes.com/2017/11/13/us/puerto-rico-hurricane-maria-mental
-health.html.

Harrington, Samantha. 2017. "Georgia Island Confronts 'Blue Sky' Floods."
Yale Climate Connections, April 21. Accessed December 26, 2017. https://
www.yaleclimateconnections.org/2017/04/georgia-island-confronts-blue
-sky-floods/.

IPCC. 2014. *Climate Change 2014: Impacts, Adaptation, and Vulnerability. Part A:
Global and Sectoral Aspects; Contribution of Working Group II to the Fifth As-
sessment Report of the Intergovernmental Panel on Climate Change,* edited by
C. B. Field, V. R. Barros, D. J. Dokken, K. J. Mach, M. D. Mastrandrea, T. E.
Bilir, M. Chatterjee, K. L. Ebi, Y. O. Estrada, R. C. Genova, B. Girma, E. S. Kis-
sel, A. N. Levy, S. MacCracken, P. R. Mastrandrea, and L. L. White. Cambridge:
Cambridge University Press. Accessed February 14, 2018. http://www.ipcc.ch
/report/ar5/wg2/.

Jula, Megan. 2018. "Hurricanes Drove More Than 1,000 Medical Evacuees from
the Virgin Islands. Many Can't Go Home." *Grist,* June 3. Accessed June 12, 2018.
https://grist.org/article/hurricanes-drove-more-than-1000-medical-evacuees
-from-the-virgin-islands-many-cant-go-home/.

Kishore, Nishant, Domingo Marqués, Ayesha Mahmud, Mathew V. Kiang,
Irmary Rodriguez, Arlan Fuller, Peggy Ebner, Cecilia Sorensen, Fabio Racy,
Jay Lemery, Leslie Maas, Jennifer Leaning, Rafael A. Irizarry, Satchit Balsari,
and Caroline O. Buckee. 2018. "Mortality in Puerto Rico after Hurricane Maria." *New England Journal of Medicine,* May 29. Accessed June 19, 2018. https://
doi.org/10.1056/NEJMsa1803972. https://www.nejm.org/doi/full/10.1056
/NEJMsa1803972.

Luhn, Alec. 2016. "Russia: Anthrax Outbreak Triggered by Climate Change
Kills Boy in Arctic Circle." *Guardian,* August 1. Accessed December 29, 2018.
https://www.theguardian.com/world/2016/aug/01/anthrax-outbreak-climate
-change-arctic-circle-russia.

Melillo, Jerry M., Terese (T. C.) Richmond, and Gary W. Yohe, eds. 2014. *Climate
Change Impacts in the United States: The Third National Climate Assessment.*
U.S. Global Change Research Program. Accessed June 22, 2017. https://doi
.org/10.7930/J0Z31WJ2.

Moshtaghian, Artemis, and Marilia Brocchetto. 2017. "This Homeowner Found
a 10-foot Gator in His Flooded Home near Houston." CNN, September 2. Accessed December 30, 2018. https://www.cnn.com/2017/09/02/weather/hurricane
-harvey-gator-homes/.

Nasci, Roger S., and Chester G. Moore. 1998. "Vector-Borne Disease Surveillance
and Natural Disasters." *Emerging Infectious Diseases* 4 (2): 333–34. Accessed
June 22, 2017. https://wwwnc.cdc.gov/eid/article/4/2/98-0227_article. https://
dx.doi.org/10.3201/eid0402.980227.

Pilkey, Orrin H., and J. Andrew G. Cooper. 2014. *The Last Beach*. Durham, NC: Duke University Press.

"Puerto Rico in Crisis as It Struggles to Treat Chronic Kidney Disease and Dialysis Patients." 2018. KidneyBuzz. Accessed March 1, 2018. https://www.kidney buzz.com/puerto-rico-in-crisis-as-it-struggles-to-treat-chronic-kidney-disease -and-dialysis-patients/.

Simmons, Daisy. 2018. "After Weather Disasters, Mental Health Problems Spike: Audio." *Yale Climate Connections*, March 20. Accessed April 5, 2018. https:// www.yaleclimateconnections.org/2018/03/after-weather-disasters-mental -health-problems-spike/.

Sneider, Noah. 2018. "Letter from Siberia. Cursed Fields: What the Tundra Has in Store for Russia's Reindeer Herders." *Harper's Magazine*, April. Accessed December 29, 2018. https://harpers.org/archive/2018/04/cursed-fields/7/.

University of Miami Miller School of Medicine. 2016. "Health Issues Due to Sea Level Rise Impact Communities in South Florida." *Science Daily*, July 12. Accessed August 16, 2016. https://www.sciencedaily.com/releases/2016/07 /160712093504.htm.

USGCRP. 2016. "Climate and Health Assessment: Water-Related Illness." In *The Impacts of Climate Change on Human Health in the United States: A Scientific Assessment.* Washington, DC: U.S. Global Change Research Program. Accessed June 22, 2017. https://health2016.globalchange.gov/.

USGCRP. 2016. *The Impacts of Climate Change on Human Health in the United States: A Scientific Assessment*, edited by A. Crimmins, J. Balbus, J. L. Gamble, C. B. Beard, J. E. Bell, D. Dodgen, R. J. Eisen, N. Fann, M. D. Hawkins, S. C. Herring, L. Jantarasami, D. M. Mills, S. Saha, M. C. Sarofim, J. Trtanj, and L. Ziska. Washington, DC: U.S. Global Change Research Program. Accessed June 22, 2017. http://dx.doi.org/10.7930/J0R49NQX.

Walsh, Mary Williams. 2017. "Puerto Rico's Finances Add to Vulnerability in Hurricane." *New York Times*, September 6. Accessed March 3, 2018. A version of this article appeared in print, September 7, 2017, page B4 of the New York edition with the headline "Puerto Rico's Troubled Finances Add to Vulnerability During Hurricane." https://www.nytimes.com/2017/09/06/business/dealbook /puerto-rico-finances-irma.html.

World Health Organization. 2014. "Climate Change 2014: Impacts, Adaptation, and Vulnerability." Report from the Intergovernmental Panel on Climate Change (IPCC). Accessed August 13, 2015. http://www.who.int/globalchange /environment/climatechange-2014-report/en/.

Chapter 5. The Front Line in the Battle

Center for Climate and Energy Solutions. 2009. "National Security Implications of Global Climate Change," August. Arlington, VA. Accessed December 28, 2017. https://www.c2es.org/site/assets/uploads/2009/08/national-security -implications-global-climate-change.pdf.

Center for Climate and Security. 2016. *Sea Level Rise and the U.S. Military's Mission: Military Expert Panel Report*, September. Accessed January 6, 2017. https://climateandsecurity.files.wordpress.com/2016/09/center-for-climate -and-security_military-expert-panel-report2.pdf.

CNA Corporation. 2007. *National Security and the Threat of Climate Change.* Accessed June 15, 2017. https://www.cna.org/cna_files/pdf/national security and the threat of climate change.pdf.

Coats, Daniel R. 2018. *Statement for the Record: Worldwide Threat Assessment of the U.S. Intelligence Community*, February 13. Washington, DC: U.S. Office of the Director of National Intelligence. Accessed March 1, 2018. https://www.dni .gov/files/documents/Newsroom/Testimonies/2018-ATA—Unclassified-SSCI .pdf.

Conca, James. 2014. "Does Our Military Know Something We Don't about Global Warming?" *Forbes*, November 14. Accessed June 25, 2018. https://www.forbest .com/sites/jamesconca/2014/11/14/does-our-military-know-something-we -dont-about-global-warming/.

Freedman, Andrew. 2014. "Virginia Officials Accomplish the Impossible: A Bipartisan Sea Level Rise Discussion." *Mashable.com*, June 30. Accessed July 25, 2017. https://mashable.com/2014/06/30/virginia-officials-accomplish-the-impossible -a-bipartisan-sea-level-rise-discussion/#pox5Nd_eWsqb.

Harrison, Don. 2015. "The Water Is Coming." *Coastal Virginia Magazine*, January. Accessed July 25, 2017. http://www.coastalvirginiamag.com/January-2015 /The-Water-Is-Coming/.

Hellegers, Desiree. 2016. "Climate Change Catch 22: It's Official—the U.S. Military Poses a Significant Threat to the U.S. Military." *CounterPunch*, September 19. Accessed October 6, 2017. https://www.counterpunch.org/2016/09/19/climate -change-catch-22-its-official-the-u-s-military-poses-a-significant-threat-to-the -u-s-military/.

Kimberlin, Joanne. 2016. "Sea Level Rise a Big Issue for Military in Hampton Roads, Science Says. But Republicans Try to Block Planning." *Virginian-Pilot*, July 27. Accessed October 6, 2017. http://pilotonline.com/news/military/local /sea-level-rise-a-big-issue-for-military-in-hampton/article_ac679bca-9972-5e50 -9f7e-b727716aa1d9.html.

Kusnetz, Nicholas. 2017. "Protecting Norfolk from Flooding Won't Be Cheap: Army Corps Releases Its Plan." *InsideClimate News*, October 30. Accessed No-

vember 2, 2017. https://insideclimatenews.org/news/30102017/norfolk-sea-level
-rising-flood-protection-plan-army-corps-engineers-climate-change.

Kusnetz, Nicholas. 2017. "Rising Seas Are Flooding Virginia's Naval Base, and
There's No Plan to Fix It." *InsideClimate News*, October 25. Accessed November
2, 2017. https://insideclimatenews.org/news/10252017/military-norfolk-naval
-base-flooding-climate-change-sea-level-global-warming-virginia.

McKenna, Phil. 2018. "Hurricane Michael Cost This Military Base About $5 Bil-
lion, Just One of 2018's Weather Disasters." *Inside Climate News*, December 18.
Accessed December 27, 2018. https://insideclimatenews.org/news/18122018
/tyndall-military-hurricane-cost-2018-year-review-billion-dollar-disasters
-wildfire-extreme-weather-drought-michael-florence.

Milman, Oliver, Emily Holden, and David Agren. 2018. "Migration: The Unseen
Driver behind the Migrant Caravan: Climate Change." *Guardian*, October 30.
Accessed December 26, 2018. https://www.theguardian.com/world/2018/oct/30
/migrant-caravan-causes-climate-change-central-america.

Myers, Meghann. 2016. "Rising Oceans Threaten to Submerge 128 Military Bases:
Report." *Navy Times*, July 29. Accessed October 26, 2016. https://www.navy
times.com/story/military/2016/07/29/rising-oceans-threaten-submerge-18
-military-bases-report/87657780/.

National Intelligence Council. 2016. *Implications for U.S. National Security of
Anticipated Climate Change: Memorandum*, September 21. NIC WP 2016–01.
Accessed November 1, 2017. https://www.dni.gov/files/documents/Newsroom
/Reports%20and%20Pubs/Implications_for_US_National_Security_of
_Anticipated_Climate_Change.pdf.

O'Connor, Mary Catherine. 2017. "Sea-Level Rise, but No Mention of 'Climate
Change.'" Yale Center for Environmental Communication: *Yale Climate Connections*, May 17. Accessed November 2, 2017. https://www.yaleclimateconnections
.org/2017/05/tidewater-sea-level-rise-minus-the-climate-change-baggage/.

Pilkey, Orrin H., and Katherine L. Dixon. 1996. *The Corps and the Shore*. Washington, DC: Island Press.

Silverstein, Ken. 2016. "Military Officials: Climate Change Could Ensnare Nation
in Humanitarian Issues and Armed Conflicts." *Forbes*, September 19. Accessed
October 2, 2016. http://www.forbes.com/sites/kensilverstein/2016/09/19
/military-officials-climate-change-could-ensnare-nation-in-humanitarian
-issues-and-armed-conflicts/—db0a4e378eee.

Spanger-Siegfried, Erika, Kristina Dahl, Astrid Caldas, Shana Udvardy, Sarah
Pendergast, Alyssa Tsuchiya, and Pamela Worth. 2016. "The U.S. Military on
the Front Lines of Rising Seas (2016)." Union of Concerned Scientists. Accessed
June 25, 2018. http://www.ucsusa.org/global-warming/global-warming-impacts
/sea-level-rise-flooding-us-military-bases#.WgH9IoZrxz8.

Titley, David. 2016. *New Dept. of Defense Directive on Climate and Security*. Center

for Climate and Security. January 20. Accessed January 7, 2018. https://climate
andsecurity.org/2016/01/20/new-dept-of-defense-directive-on-climate-security/.

Union of Concerned Scientists. 2016. "Land Loss across Bases." Accessed March 8,
2018. http://www.ucsusa.org/sites/default/files/images/2016/07/gw-impacts
-military-chart-all-bases.jpg.

U.S. Department of Defense. 2016. *DoD Directive 4715.21. Climate Change Adaptation and Resilience*, January 14. Washington, DC: U.S. Office of the Under Secretary of Defense for Acquisition, Technology, and Logistics. Accessed February
15, 2018. https://www.defense.gov/Portals/1/Documents/pubs/471521p.pdf.

U.S. Department of Defense. 2018. "About Us: Joint Base Charleston." Accessed
January 7, 2018. http://www.jbcharleston.jb.mil/About-Us/.

Chapter 6. At-Risk Coastal Environments

Barnard, Patrick L., Daniel Hoover, David M. Hubbard, Alex Snyder, Bonnie C.
Ludka, Jonathan Allan, George M. Kaminsky, Peter Ruggiero, Timu W.
Gallien, Laura Gabel, Diana McCandless, Heather M. Weiner, Nicholas Cohn,
Dylan L. Anderson, and Katherine A. Serafin. 2017. "Extreme Oceanographic
Forcing and Coastal Response Due to the 2015–2016 El Niño." *Nature Com-
munications* 8, 14365. Accessed February 15, 2018. https://doi.org/10.1038
/ncomms14365 (2017).

Beavers, Rebecca, Amanda Babson, and Courtney Schupp, eds. 2016. *Coastal Adaptation Strategies Handbook*. NPS 999/134090. Washington, DC: National Park
Service. Accessed October 12, 2017. https://www.nps.gov/subjects/climate
change/coastalhandbook.htm

Carroll, Rebecca. 2009. "Mississippi River Delta to 'Drown' by 2100?" *National
Geographic News*, June 29. Accessed October 1, 2017. https://news.national
geographic.com/news/2009/06/090629-mississippi-river-sea-levels.html.

Chapman, Alex, and Van Pham Dang Tri. 2018. "Climate Change Is Triggering
a Migrant Crisis in Vietnam." *Conversation*, January 9. Accessed May 25, 2018.
https://theconversation.com/climate-change-is-triggering-a-migrant
-crisis-in-vietnam-88791.

Daugherty, Alex, and Joey Flechas. 2017. "Replacing Miami's Beach Sands Costs
Millions. Here's How Congress Could Make It Cheaper." *Miami Herald*, October 24. Accessed November 2, 2017. http://www.miamiherald.com/news
/politics-government/article180602091.html.

Feagin, Rusty A., Douglas J. Sherman, and William E. Grant. 2005. "Coastal
Erosion, Global Sea-Level Rise, and the Loss of Sand Dune Plant Habitats."
Frontiers in Ecology and the Environment 3 (7): 359–64. Accessed November 2,
2017. Published by Wiley on behalf of the Ecological Society of America.
https://doi.org/10.2307/3868584.

Flavelle, Christopher. 2018. "As Storms Get Stronger, Building Codes Are Getting

Weaker." Bloomberg, March 19. Accessed May 22, 2018. https://www
.bloomberg.com/news/articles/2018–03–19/storm-prone-states-ease-off
-building-codes-as-climate-risk-grows.

Grabar, Henry. 2017. "South Florida, Out of Beach, Wants to Buy Sand from the
Bahamas." *Moneybox: Slate.com*, November 2. Accessed November 4, 2017.
https://slate.com/business/2017/11/south-floridas-beaches-are-disappearing
-and-the-state-wants-to-import-sand-from-the-bahamas.html.

Janin, Hunt, and Scott A. Mandia. 2012. *Rising Sea Levels: An Introduction to
Cause and Impact.* Jefferson, NC: McFarland.

Kim, Oanh Le Thi, and Truong Le Minh. 2017. "Correlation between Climate
Change Impacts and Migration Decisions in Vietnamese Mekong Delta."
IJISET—International Journal of Innovative Science, Engineering and Technology
4 (8): 111–18. Accessed June 19, 2018. ISSN (Online) 2348—7968 | Impact Factor
(2016)—5.264. http://ijiset.com/vol4/v4s8/IJISET_V4_I08_13.pdf.

Limber, Patrick W., Patrick L. Barnard, Sean Vitousek, and Li H. Erikson. 2018.
"A Model Ensemble for Projecting Multidecadal Coastal Cliff Retreat during
the 21st Century." *Journal of Geophysical Research: Earth Surface* 6 (3). Accessed
July 4, 2018. https://doi.org/10.1029/2017JF004401.

Milman, Oliver. 2017. "'Buried in Marshes': Sea-Level Rise Could Destroy Historic
Sites on U.S. East Coast." *Guardian*, November 27. Accessed December 15,
2017. https://www.theguardian.com/environment/2017/nov/29/buried-in
-marshes-sea-level-rise-could-destroy-historic-sites-on-us-east-coast.

Peterson, Bo. 2017. "First Inland South Carolina Tract Purchased in Cape Romain
Effort to Save Habitat as Seas Rise." *Post and Courier*, November 24. Accessed
December 14, 2017. https://www.google.com/search?q=First+inland+South
+Carolina+tract+purchased+in+Cape+Romain+effort+to+save+habitat+as
+seas+rise&ie=utf-8&oe=utf-8&client=firefox-b-1-ab.

Pilkey, Orrin H., and Tom W. Davis. 1987. "An Analysis of Coastal Recession Mod-
els: North Carolina Coast." In "Sea Level Rise and Coastal Evolution," edited by
D. Nummedal, O. H. Pilkey, and J. D. Howard. *Society of Economic Paleontolo-
gists and Mineralogists* (SEPM) Special Publication no. 41: 59–68.

Pilkey, Orrin H., with original batiks by Mary Edna Fraser. 2003. *A Celebration of
the World's Barrier Islands.* New York: Columbia University Press.

Pilkey, Orrin H., William J. Neal, Joseph T. Kelley, and J. Andrew G. Cooper. 2011.
The World's Beaches: A Global Guide to the Science of the Shoreline. Berkeley:
University of California Press.

Pilkey, Orrin H., and Howard L. Wright III. 1988. "Seawalls versus Beaches." In
"The Effects of Seawalls on the Beach," edited by N. C. Kraus and O. H. Pilkey.
Journal of Coastal Research, Special Issue no. 4: 41–64. https://www.jstor.org
/stable/25735351.

Program for the Study of Developed Shorelines. 2018. "Beach Nourishment Viewer."
Cullowee, NC: Western Carolina University. http://beachnourishment.wcu.edu/.

Rush, Elizabeth. 2015. "Down on the Disappearing Bayou." *Le Monde diplomatique*, November. Accessed October 6, 2017. http://mondediplo.com/2015/11/1110uisiana.

Stutz, Matthew L., and Orrin H. Pilkey. 2001. "A Review of Global Barrier Island Distribution." ICS 2000 Proceedings. *Journal of Coastal Research*, Special Issue no. 34: 15–22.

Sweeney, Dan. 2018. "Florida Has Spent More Than $100 Million Pouring More Sand onto Beaches in the Past Three Years. Is It Time to Wave a White Flag?" *South Florida Sun Sentinel*, June 8. Accessed June 25, 2018. http://www.sun-sentinel.com/news/sound-off-south-florida/fl-reg-beach-renourishment-20180608-story.html.

Union of Concerned Scientists. 2018. "Underwater: Rising Seas, Chronic Floods, and the Implications for U.S. Coastal Real Estate." June. Accessed June 24, 2018. https://www.ucsusa.org/sites/default/files/attach/2018/06/underwater-analysis-full-report.pdf.

Van Alstyne, Lewis III. 2016. "Changing Winds and Rising Tides on Beach Renourishment in Florida: Short-Term Alternatives and Long-Term Sustainable Solutions Using Law and Policy from Florida and Nearby States." *Florida A&M University Law Review* 11, no. 2, article 5. http://commons.law.famu.edu/famulawreview/vol11/iss2/5.

Xia, Rosanna. 2018. "Southern California's Coastal Communities Could Lose 130 Feet of Cliffs This Century as Sea Levels Rise." *Los Angeles Times,* June 27. Accessed July 2, 2018. http://www.latimes.com/local/lanow/la-me-cliff-erosion-sea-level-rise-20180627-story.html.

Chapter 7. The Environmental Impact of Surging Seas

Blum, Linda J. 2011. "Coastal Marshes and Rising Sea Levels." University of Virginia, Virginia Coast Reserve/Long-Term Ecological Research Program (VCR/LTER). Accessed October 25, 2017. https://lternet.edu/research/keyfindings/coastal-marshes-and-rising-sea-levels.

Center for Biological Diversity. 2013. "Deadly Waters: How Rising Seas Threaten 233 Endangered Species." *Sea Level Rise Report,* December. Accessed November 2, 2017. https://www.biologicaldiversity.org/campaigns/sea-level_rise/pdfs/SeaLevelRiseReport_2013_print.pdf.

Feagin, Rusty A., Douglas J. Sherman, and William E. Grant. 2005. "Coastal Erosion, Global Sea-Level Rise, and the Loss of Sand Dune Plant Habitats." *Frontiers in Ecology and the Environment* 3 (7): 359–64.

Fox, Alex. 2017. "Threatened Sea Turtles in Hawaii Losing Ground to Rising Oceans." *EOS* 98, December 14. Accessed January 16, 2018. https://doi.org/10.1029/2017EO089065.

Garner, Kendra L., Michelle Y. Chang, Matthew T. Fulda, Jonathan A. Berlin, Rachel E. Freed, Melissa M. Soo-Hoo, Dave L. Revell, Makihiko Ikegami, Lorraine E. Flint, Alan L. Flint, and Bruce E. Kendall. 2015. "Impacts of Sea Level Rise and Climate Change on Coastal Plant Species in the Central California Coast." *Peer J* 3: e958. Accessed June 25, 2018. https://doi.org/10.7717/peerj.958.

Jacobi, James. 2016. "Impact of Sea Level Rise on Coastal Plants and Cultural Sites." Pacific Island Ecosystems Research Center, U.S. Geological Survey. Accessed November 2, 2017. https://www.usgs.gov/centers/pierc/science/impact-sea-level-rise-coastal-plants-and-cultural-sites?qt-science_center_objects=0#qt-science_center_objects.

Norwegian Institute of Bioeconomy Research. 2017. "Food Security Threatened by Sea-Level Rise." Phys.org, January 18. Accessed November 2, 2017. https://phys.org/news/2017-01-food-threatened-sea-level.html.

Peterson, Charles H., and Melanie J. Bishop. 2005. "Assessing the Environmental Impacts of Beach Nourishment." *BioScience* 55, no. 10 (October 1): 887–96. Accessed November 5, 2016. https://doi.org/10.1641/0006-568(2005)055[0887:ATEIOB]2.0.CO;2.

Reed, Denise J. 1990. "The Impact of Sea-Level Rise on Coastal Salt Marshes." *Progress in Physical Geography* 14 (4) (December 1): 465–81. Accessed November 2, 2017. https://doi.org/10.1177/030913339001400403.

Rice, Doyle. 2013. "Sea-Level Rise Threatens Hundreds of U.S. Animal Species." *USA Today*, December 11. Updated December 11, 2013. Accessed March 3, 2017. https://www.usatoday.com/story/weather/2013/12/11/sea-level-rise-threatened-endangered-species/3963339/.

Union of Concerned Scientists. 2017. "North Carolina Faces Chronic Inundation: Fact Sheet." Accessed November 2, 2017. http://www.ucsusa.org/RisingSeasHitHome.

Chapter 8. Inundated Infrastructure

Azevedo de Almeida, Beatriz, and Ali Mostafavi. 2016. "Resilience of Infrastructure Systems to Sea-Level Rise in Coastal Areas: Impacts, Adaptation Measures, and Implementation Challenges." *Sustainability* 8, no. 11 (November 1): 1115. Accessed March 18, 2018. https://doi.org/10.3390/su8111115.

Behr, Peter. 2014. "NUCLEAR: Violations at Fla. Plant Highlight NRC Concerns Over Flooding Emergencies." *E&E News: Environment and Energy Publishing*, December 1. Accessed March 18, 2018. https://www.eenews.net/stories/1060009671.

Bovarnick, Ben, Shiva Polefka, and Arpita Bhattacharyya. 2014. "Rising Waters, Rising Threat: How Climate Change Endangers America's Neglected Wastewater Infrastructure," October 31. Washington, DC: Center for American Progress. Accessed April 13, 2018. https://cdn.americanprogress.org/wp-content/uploads/2014/10/wastewater-report.pdf.

Chin, David. 2015. "The Cooling-Canal System at the FPL Turkey Point Power Station." Miami-Dade County and the University of Miami, September 15. Accessed March 25, 2018. http://www.miamidade.gov/environment/library/reports/cooling-canal-system-at-the-fpl-turkey-point-power-station.pdf.

Dearen, Jason, Michael Biesecker, and Angeliki Kastanis. 2017. "AP Finds Climate Change Risk for 327 Superfund Toxic Sites." Associated Press AP News, December 22. Accessed April 14, 2018. https://apnews.com/31765cc6d10244588805ee738edcb36b/AP-finds-climate-change-risk-for-327-toxic-Superfund-sites?utm_source=Daily+Climate&utm_campaign=1fd4daaaf1-RSS_EMAIL_CAMPAIGN&utm_medium=email&utm_term=0_9c8bdfd977-1fd4daaaf1-99370265.

Dorfman, Mark, Nancy Stoner, and Michele Merkel. 2004. "Swimming in Sewage: The Growing Problem of Sewage Pollution and How the Bush Administration Is Putting Our Health and Environment at Risk." Natural Resources Defense Council and the Environmental Integrity Project, February. Accessed March 25, 2018. http://wcsu.csu.edu/cerc/documents/SwimmingInSewage.pdf.

Durairajan, Ramakrishnan, Carol Barford, and Paul Barford. 2018. "Lights Out: Climate Change Risk to Internet Infrastructure." Applied Networking Research Workshop (ANRW '18), Montreal, QC, Canada, July 16. Accessed July 20, 2018. https://doi.org/10.1145/3232755.3232775.

Flavelle, Christopher, and Jeremy C. F. Lin. 2018. "Rising Waters Are Drowning Amtrak's Northeast Corridor." Bloomberg *Businessweek*. December 20. Accessed December 22, 2018. https://www.bloomberg.com/graphics/2018-amtrak-sea-level/.

Flechas, Joey. 2017. "Miami Beach to Begin $100 Million Flood Prevention Project in Face of Sea Level Rise." *Miami Herald*, January 28. Accessed October 25, 2017. http://www.miamiherald.com/news/local/community/miami-dade/miami-beach/article129284119.html.

Fox, Alex. 2017. "Sea Level Rise May Swamp Many Coastal U.S. Sewage Plants." *EOS, Earth and Space Science News*, December 13. Accessed April 13, 2018. https://eos.org/articles/sea-level-rise-may-swamp-many-coastal-u-s-sewage-plants.

Given, Suzan, Linwood H. Pendleton, and Alexandria B. Boehm. 2006. "Regional Public Health Cost Estimates of Contaminated Coastal Waters: A Case Study of Gastroenteritis at Southern California Beaches." *Environmental Science and Technology* 40 (16): 4851–58. Accessed March 24, 2018. https://pubs.acs.org/doi/pdf/10.1021/es060679s.

Harris, Alex. 2018. "Keys to Raise Roads before Climate Change Puts Them

Underwater. It'll Be Expensive." *Miami Herald*, February 1. Accessed April 13, 2018. http://www.miamiherald.com/news/local/environment/article197735369 .html.

Hesterman, Donna. 2011. "Seaports Need a Plan for Weathering Climate Change, Say Stanford Researchers." *Stanford News Blog*, May 6. Accessed April 13, 2018. https://news.stanford.edu/news/2011/may/seaports-climate-change-051611 .html.

Hummel, Michelle A., Matthew S. Berry, and Mark T. Stacey. 2018. "Sea Level Rise Impacts on Wastewater Treatment Systems along the U.S. Coasts." *Earth's Future: An Open Access AGU Journal*, March 24. Accessed April 13, 2018. https:// doi.org/10.1002/2017EF000805. https://agupubs.onlinelibrary.wiley.com/doi /pdf/10.1002/2017EF000805.

Iannelli, Jerry. 2017. "State Senator Says FPL Isn't Preparing Miami's Nuclear Plant for Sea-Level Rise." *Miami New Times*, October 5. Accessed March 18, 2018. http://www.miaminewtimes.com/news/turkey-point-miami-nuclear-plant-sea -level-rise-plan-inadequate-miami-lawmaker-warns-9722390.

Lerner, Sharon. 2018. "Hurricane Florence Released Tons of Coal Ash in North Carolina. Now the Coal Industry Wants Less Regulation." *The Intercept*, September 28. Accessed December 27, 2018. https://theintercept.com/2018/09/28 /north-carolina-coal-ash-hurricane-florence/.

Lochbaum, Dave. 2011. "Fission Stories #48: Hurricane Andrew vs. Turkey Point." Union of Concerned Scientists: *Blog, All Things Nuclea*, July 12. Accessed March 18, 2018. http://allthingsnuclear.org/dlochbaum/fission-stories-48 -hurricane-andrew-vs-turkey-point.

McGeehan, Patrick, and Winnie Hu. 2017. "Five Years after Sandy, Are We Better Prepared?" *New York Times*, October 29. A version of this article appeared in print on October 30, 2017, on page A1 of the New York edition with the headline "5 Years after Storm Surge, Big Plans and Unfinished Protections." Accessed March 25, 2018. https://www.nytimes.com/2017/10/29/nyregion/five-years-after -sandy-are-we-better-prepared.html.

"Motiva Shuts Port Arthur Texas Refinery Due to Flooding." 2017. *Energy: CNBC. com*, August 30. Accessed March 15, 2018. https://www.cnbc.com/2017/08/30 /motiva-shuts-port-arthur-texas-refinery-due-to-flooding.html.

Nunez, Christina. 2015. "As Sea Levels Rise, Are Coastal Nuclear Plants Ready?" *National Geographic*, December 16. Accessed August 30, 2017. https://news .nationalgeographic.com/energy/2015/12/151215-as-sea-levels-rise-are -coastal-nuclear-plants-ready/.

Robinson, Kim Stanley. 2017. *New York 2140*. London: Orbit.

Rodríguez, José Javier (Senator, The Florida Senate, District 37, Miami). 2017. Letter to Honorable Kristine L. Svinicki (Chairman, United States Nuclear Regulatory Commission). August 23. 2 leaves. Sent via email to chairman@nrc.gov and

hearingdocket@nrc.gov. Accessed January 7, 2019. https://www.nrc.gov/docs
/ML1723/ML17235B122.pdf.

Rosenzweig, Cynthia, William D. Solecki, Reginald Blake, Malcolm Bowman, Craig Faris, Vivien Gornitz, Radley Horton, Klaus Jacob, Alice LeBlanc, Robin Leichenko, Megan Linkin, David Major, Megan O'Grady, Lesley Patrick, Edna Sussman, Gary Yohe, and Rae Zimmerman. 2011. "Developing Coastal Adaptation to Climate Change in the New York City Infrastructure-Shed: Process, Approach, Tools, and Strategies." *Climatic Change* 106: 93–127. Accessed March 25, 2018. https://doi.org/10.1007/s10584–010–0002–8.

Salisbury, Susan. 2017. "Groups Win Round in Court Over FPL's Turkey Point Pollution." Palm Beach Post, November 28. Accessed March 25, 2018. https://www.palmbeachpost.com/business/groups-win-round-court-over-fpl -turkey-point-pollution/Vcv6dUnoygjrOOrxNdS70K/.

Salisbury, Susan. 2017. "PSC: FPL Can Charge Customers for Turkey Point Cooling Canal Fix." PalmBeachPost.com, December 12. Accessed January 5, 2018. https:// www.palmbeachpost.com/business/psc-fpl-can-charge-customers-for-turkey -point-cooling-canal-fix/RK7UgPrRA6y8GjYMaXEH5J/amp.html.

Satterfield, Jamie. 2018. "Jury: Jacobs Engineering Endangered Kingston Disaster Clean-up Workers," November 7. Updated November 8. *Knox News.* Accessed December 27, 2018. https://www.knoxnews.com/story/news/crime/2018/11/07 /verdict-reached-favor-sickened-workers-coal-ash-cleanup-lawsuit/1917514002/.

Saunders, Jim. 2017. "FPL Gets State's OK to Charge Customers for Cleaning Saltwater Plume at Turkey Point." *Miami Herald*, December 12. Accessed April 7, 2018. http://www.miamiherald.com/news/local/community/miami-dade /article189398169.html.

Song, Lisa. 2011. "Decommissioning a Nuclear Plant Can Cost $1 Billion and Take Decades." *Reuters: SolveClimate News,* June 13. Accessed March 25, 2018. https:// www.reuters.com/article/idUS178883596820110613.

Strauss, Ben, and Remik Ziemlinski. 2012. "Sea Level Rise Threats to Energy Infrastructure: A Surging Seas Brief Report by Climate Central." *Climate Central,* April 19. Accessed March 25, 2018. http://slr.s3.amazonaws.com/SLR-Threats-to -Energy-Infrastructure.pdf.

Tabuchi, Hiroko, Nadja Popovich, Blacki Migliozzi, and Andrew W. Lehren. 2018. "Floods Are Getting Worse, and 2,500 Chemical Sites Lie in the Water's Path." *New York Times*, February 6. Accessed April 14, 2018. https://www.nytimes .com/interactive/2018/02/06/climate/flood-toxic-chemicals.html.

U.S. Nuclear Regulatory Commission. 2014. "Escalated Enforcement Actions Issued to Reactor Licensees—S: Saint Lucie 1 & 2." Docket Nos. 050-00335; 050 -00389. EA-14-131, November 19. Accessed January 7, 2019. https://www.nrc.gov /reading-rm/doc-collections/enforcement/actions/reactors/s.html#SaintLucie.

Wald, Matthew L. 2012. "Heat Shuts Down a Nuclear Reactor." *NYTimes Green-*

blog, August 31. Accessed March 18, 2018. https://green.blogs.nytimes.com
/2012/08/13/heat-shuts-down-a-coastal-reactor/.

Wang, Ruo-Qian, Mark T. Stacey, Liv Herdman, Patrick Barnard, and Li Erik-
son. 2018. "The Influence of Sea Level Rise on the Regional Interdependence of
Coastal Infrastructure." *Earth's Future: An Open Access AGU Journal*, March 25.
Accessed April 5, 2018. https://doi.org/10.1002/2017EF000742.

Wenner, Elizabeth, Denise Sanger, Saundra Upchurch, and M. Thompson. "Char-
acterization of the Ashepoo-Combahee-Edisto (ACE) Basin, South Carolina."
South Carolina Department of Natural Resources (SCDNR) Marine Resources
Research Institute. n.d. Accessed July 3, 2018. https://webapp1.dlib.indiana.edu
/virtual_disk_library/index.cgi/4928836/FID1596/intro.htm.

Chapter 9. Coastal Catastrophes

99% Invisible. 2018. "Managed Retreat, Episode 293." Podcast, January 30.
Accessed February 22, 2018. https://99percentinvisible.org/episode/managed
-retreat/.

Al, Stefan. 2018. *Adapting Cities to Sea Level Rise: Green and Gray Strategies.*
Washington, DC: Island Press.

Brey, Jared. 2018. "Hawaii Gets Explicit about Sea-Level Rise." *Next City*, June 14.
Accessed December 30. 2018. https://nextcity.org/daily/entry/hawaii-gets
-explicit-about-sea-level-rise.

Buchanan, Maya K., Michael Oppenheimer, and Robert E. Kopp. 2017. "Amplifica-
tion of Flood Frequencies with Local Sea Level Rise and Emerging Flood Re-
gimes." *Environmental Research Letters* 12 (6): 064009. Accessed June 17, 2019.
https://iopscience.iop.org/article/10.1088/1748-9326/aa6cb3/meta.

Burch, Audra D. S. 2018. "Brutal Choice in Houston: Sell Home at a Loss or Face
New Floods." *New York Times*, March 30. Accessed April 3, 2018. https://www
.nytimes.com/2018/03/30/us/hurricane-harvey-flooding-canyon-gate.html.

Carroll, Susan, and Mike Ward. 2018. "A Massive Seawall for Southeast Texas
Could Save Money. But So Far, There's None to Spend." *Houston Chronicle*,
January 5. Updated January 6, 2018. Accessed January 26, 2018. http://www
.houstonchronicle.com/news/houston-texas/houston/article/A-massive-seawall
-for-Southeast-Texas-could-save-12477716.php.

Chassignet, Eric P., James W. Jones, Vasubandhu Misra, and Jayantha Obeysekera,
eds. 2017. *Florida's Climate: Changes, Variations, and Impacts.* Gainesville:
Florida Climate Institute. Accessed January 16, 2018. https://doi.org/10.17125
/fci2017.

Dawson, Ashley. 2017. *Extreme Cities: The Peril and Promise of Urban Life in the
Age of Climate Change.* New York: Verso.

Drew, James. 2018. "A Fort Bend Engineer's Warning, 25 Years Old, Comes True
during Harvey." *Houston Chronicle*, January 20. Updated January 22, 2018. Ac-

cessed March 15, 2018. https://www.houstonchronicle.com/news/houston
-texas/houston/article/A-Fort-Bend-engineer-s-warning-25-years-old-12511632
.php?t=11b9850018.

Evans, Beau. 2018. "46 Tons of Mardi Gras Beads Found in Clogged Catch Basins."
NOLA: The Times-Picayune, January 25. Accessed February 3, 2018. http://www
.nola.com/politics/index.ssf/2018/01/catch_basins_cleaned_mardi_gra.html.

Fears, Darryl. 2017. "Tampa Bay's Coming Storm." Washington Post, July 28. Accessed April 26, 2018. https://www.washingtonpost.com/graphics/2017/health
/environment/tampa-bay-climate-change/?noredirect=on.

Flavelle, Christopher. 2018. "The Fighting Has Begun over Who Owns Land
Drowned by Climate Change." Bloomberg Businessweek, April 25. Accessed
June 28, 2018. https://www.bloomberg.com/news/features/2018–04–25/fight
-grows-over-who-owns-real-estate-drowned-by-climate-change.

Gold, Russell. 2018. "Boston Agonizes over How to Protect Itself from Future
Storms." Wall Street Journal, January 28. Accessed February 23, 2018.
https://www.wsj.com/articles/boston-agonizes-over-how-to-protect-itself
-from-future-storms-1517054400.

Goodell, Jeff. 2017. The Water Will Come: Rising Seas, Sinking Cities, and the Re-
making of the Civilized World. New York: Little, Brown.

Hanson, Susan, Robert J. Nicholls, Nicola Patmore, Stéphane Hallegatte, Jan
Corfee-Morlot, Celine Herweijer, and Jean Château. 2011. "A Global Ranking of
Port Cities with High Exposure to Climate Extremes." Climatic Change 140 (1):
89–111. Accessed October 28, 2017. https://doi.org/10.1007/s10584–010–9977–4.

Houston Downtown Management District (Downtown District). 2017. "Plan
Downtown: Converging Culture, Lifestyle and Commerce." Report by
downtowndistrict.org, November. Accessed January 19, 2018. http://www.down
towndistrict.org/static/media/uploads/attachments/plan_downtown_report
_final_spreads_sm.pdf.

Kimmelman, Michael. 2017. "Lessons from Hurricane Harvey: Houston's Struggle
Is America's Tale." New York Times. November 11. Accessed June 25, 2018.
https://www.nytimes.com/interactive/2017/11/11/climate/houston-flooding
-climate.html.

Kingwood Association Management. 2018. "District E March Newsletter." An-
nouncement to the Kingwood Association, Houston, TX, March 4. Accessed
March 16, 2018. http://kingwoodassociationmanagement.com/kingwoodmgt
/announcement.asp?id=55.

Kolbert, Elizabeth. 2015. "The Siege of Miami." New Yorker, December 21 and 28.
Accessed May 3, 2018. https://www.newyorker.com/magazine/2015/12/21/the
-siege-of-miami.

Krishna, Roy. 2017. "Climate Change and Housing: Will a Rising Tide Sink all
Homes?" Zillow.com. Accessed November 2, 2017. https://www.zillow.com
/research/climate-change-underwater-homes-12890/.

Kulp, Scott, Benjamin Strauss, Dyonishia Nieves, Shari Bell, and Dan Rizza. 2017. "These U.S. Cities Are Most Vulnerable to Major Coastal Flooding and Sea Level Rise." *Climate Central*, October 25. Accessed October 28, 2017. http://www.climatecentral.org/news/us-cities-most-vulnerable-major -coastal-flooding-sea-level-rise-21748.

Kusnetz, Nicholas. 2018. "Norfolk Wants to Remake Itself as Sea Level Rises, but Who Will Be Left Behind?" *InsideClimate News*. Accessed June 2, 2018. https://insideclimatenews.org/news/15052018/norfolk-virginia-navy-sea-level -rise-flooding-urban-planning-poverty-coastal-resilience?amp&__twitter _impression=true.

Larson, Erik. 2000. *Isaac's Storm: A Man, a Time, and the Deadliest Hurricane in History*. New York: Random House.

Magill, Bobby. 2015. "Katrina: Lasting Climate Lessons for a Sinking City." Climate Central. August 26. https://www.climatecentral.org/news/katrina-climate -change-sinking -ground-19370.

Marcelo, Philip. 2018. "Flooding a Threat to GE's New Boston Headquarters." Associated Press, *Hartford Courant*, June 6. Accessed January 8, 2019. https://www .courant.com/business/hc-biz-ge-headquarters-flooding-20180606-story.html.

Miami-Dade County Department of Regulatory and Economic Resources, Miami-Dade County Water and Sewer Department, and Florida Department of Health in Miami-Dade County (Dr. Samir Elmir). 2018. "Septic Systems Vulnerable to Sea Level Rise: Final Report in Support of Resolution No. R911-16," November. Accessed January 11, 2019. https://www.miamidade.gov /green/library/vulnerability-septic-systems-sea-level-rise.pdf.

Mulkern, Anne C. 2018. "Climate Impacts: Calif. Prepares Policy for Coastal 'Retreat'." *E&E News*, December 7. Accessed December 30, 2018. https://www .eenews.net/stories/1060109045.

Nicholls, Robert J., Susan Hanson, Celine Herweijer, Nicola Patmore, Stéphane Hallegatte, Jan Corfee-Morlot, Jean Château, Robert Muir-Wood. 2008. "Ranking Port Cities with High Exposure and Vulnerability to Climate Extremes." *OECD Environment Working Papers*, no. 1 (November 19). Paris: OECD. Accessed June 25, 2018. https://doi.org/10.1787/011766488208.

Olsen, Lise. 2017. "For Buyers within 'Flood Pools,' No Warnings from Developers, Public Officials." Developing Storm: Part 6. *Houston Chronicle*, December 21. Accessed January 23, 2018. http://www.houstonchronicle.com/news/houston -texas/houston/article/For-buyers-within-flood-pools-no-warnings-12434078 .php.

Plautz, Jason. 2018. "Honolulu Mayor Orders Preparations for Sea Level Rise." *Smart Cities Dive*, July 23. Accessed December 30, 2018. https://www .smartcitiesdive.com/news/honolulu-mayor-orders-preparations-for-sea -level-rise/528322/.

Ritchie, Bruce. 2017. "Scott Funding Request to Address Sea Level Rise Seen as

Turnaround for Administration." *Politico Florida*, November 20. Accessed
April 6, 2018. https://www.politico.com/states/florida/story/2017/11/20
/scott-funding-request-to-address-sea-level-rise-seen-as-turnaround-for
-administration-118725.

Rush, Elizabeth. 2018. *Rising: Dispatches from the New American Shore*. Minne-
apolis: Milkweed Editions.

Schwartz, Jen. 2018. "Surrendering to Rising Seas." *Scientific American*, August.
Accessed July 17, 2018. https://www.scientificamerican.com/article/surrendering
-to-rising-seas/.

Sullivan, Patricia. 2018. "This Coastal Town's Battle against Sea-Level Rise Could
Offer Lessons for Others." *Washington Post*, July 27. Accessed August 1,
2018. https://www.washingtonpost.com/local/virginia-politics/this-coastal-
towns-battle-against-sea-level-rise-could-offer-lessons-for-others/2018/07
/26/8c7e43a2–8b6c-11e8–85ae-511bc1146b0b_story.html?utm_term
=.a79d77dd0d24.

Tampa Bay Regional Planning Council. 2017. "The Cost of Doing Nothing: Eco-
nomic Impacts of Sea Level Rise in the Tampa Bay Area," January 21. Accessed
March 2, 2018. http://www.tbrpc.org/wp-content/uploads/2018/11/2017-The
_Cost_of_Doing_Nothing_Final.pdf.

The Citadel. 2018. "War Between the States." *War Deaths: Citadel History*. Accessed April 18, 2018. http://www.citadel.edu/citadel-history/war-deaths/war
-between-the-states.html.

Viglucci, Andres, and Joey Flechas. 2017. "South Beach Wants to Save Art Deco
Gems before the Seas Rise. One Solution: Jack Them Up." *Miami Herald*,
April 14. Accessed June 18, 2017. http://www.miamiherald.com/news/local
/community/miami-dade/miami-beach/article144575594.html.

Wong, Poh Poh, Iñigo J. Losada, Jean-Pierre Gattuso, Jochen Hinkel, Abdella-
tif Khattabi, Kathleen L. McInnes, Yoshiki Saito, and Asbury Sallenger. 2014.
"Coastal Systems and Low-lying Areas: Deltas 5.4.2.7 (380–81)." In *Climate
Change 2014: Impacts, Adaptation, and Vulnerability. Part A: Global and Sec-
toral Aspects; Contribution of Working Group II to the Fifth Assessment Report
of the Intergovernmental Panel on Climate Change*, edited by C. B. Field, V. R.
Barros, D. J. Dokken, K. J. Mach, M. D. Mastrandrea, T. E. Bilir, M. Chatterjee,
K. L. Ebi, Y. O. Estrada, R. C. Genova, B. Girma, E. S. Kissel, A. N. Levy,
S. MacCracken, P. R. Mastrandrea, and L. L. White, 361–409. Cambridge:
Cambridge University Press.

Zandonella, Catherine. 2017. "Rising Sea Levels Will Boost Moderate Floods in
Some Areas, Severe Floods in Others." Princeton University Office of the
Dean for Research. June 8. Accessed October 24, 2018. https://www.princeton
.edu/news/2017/06/08/rising-sea-levels-will-boost-moderate-floods-some-areas
-severe-floods-others.

115th Congress (2017–18). H.R. 2874—21st Century Flood Reform Act. Sponsor: Rep. Sean P. Duffy (R-WI-7). (Introduced June 12, 2017). Washington, DC. Congress.gov. Accessed January 1, 2018. https://www.congress.gov/bill/115th -congress/house-bill/2874/text.

Adams, Jennifer D., Crystal Fortwangler, and Hadiya Gibney Sewer. 2017. "Green Islands for All? Avoiding Climate Gentrification in the Caribbean." Tacoma, WA: University of Puget Sound, Society of Ethnobiology, October 19. Accessed February 1, 2018. https://ethnobiology.org/forage/blog/green -islands-all-avoiding-climate-gentrification-caribbean.

Barlow, Christine G. 2017. "The 21st Century Flood Reform Act: What's in It?" National Underwriter Property and Casualty: *Property Casualty 360*, November 15. Accessed January 1, 2018. http://www.propertycasualty360.com/2017/11/15 /the-21st-century-flood-reform-act-whats-in-it.

Bernstein, Asaf, Matthew Gustafson, and Ryan Lewis. 2018. "Disaster on the Horizon: The Price Effect of Sea Level Rise." *Elsevier: Social Science Research Network (SSRN)*, May 4. Last revised May 15, 2018. Accessed June 3, 2018. https:// ssrn.com/abstract=3073842 or https://dx.doi.org/10.2139/ssrn.3073842 or http:// leeds-faculty.colorado.edu/AsafBernstein/DisasterOnTheHorizon_PriceOfSLR _BGL.pdf.

Bolstad, Erika. 2017. "High Ground Is Becoming Hot Property as Sea Level Rises: Climate Change May Now Be a Part of the Gentrification Story in Miami Real Estate." *Scientific American*, May 1. Accessed January 31, 2018. https://www .scientificamerican.com/article/high-ground-is-becoming-hot-property-as -sea-level-rises/.

Casselman, Ben. 2015. "Katrina Washed Away New Orleans's Black Middle Class." FiveThirtyEight.com, August 24. Accessed January 31, 2018. https://fivethirty eight.com/features/katrina-washed-away-new-orleanss-black-middle-class/.

Chicago Tribune Editorial Board. 2017. "Editorial: Floods Are Inevitable. Soaking Taxpayers Isn't." *Chicago Tribune*, November 26. Accessed December 31, 2017. http://www.chicagotribune.com/news/opinion/editorials/ct-edit-flood -insurance-hurricanes-harvey-20171110-story.html.

Cleetus, Rachel. 2013. "Overwhelming Risk: Rethinking Flood Insurance in a World of Rising Seas (2013)," August. Cambridge, MA: Union of Concerned Scientists. Revised February 2014. Accessed December 31, 2017. https://www .ucsusa.org/global_warming/science_and_impacts/impacts/flood-insurance -sea-level-rise.html. https://www.ucsusa.org/sites/default/files/legacy/assets /documents/global_warming/Overwhelming-Risk-Full-Report.pdf.

Cooke, Bill. 2016. "Remembering Miami Beach's Shameful History of Segregation and Racism." *Miami New Times*, March 10. Accessed April 1, 2018. http://www

.miaminewtimes.com/news/remembering-miami-beachs-shameful-history-of
-segregation-and-racism-8306647.

Dennis, Brady. 2017. "The Country's Flood Insurance Program Is Sinking. Rescu-
ing It Won't Be Easy." *Washington Post*, July 16. Accessed December 31, 2017.
https://www.washingtonpost.com/national/health-science/the-countrys
-flood-insurance-program-is-sinking-rescuing-it-wont-be-easy/2017/07/16
/dd766c44–6291–11e7–84a1-a26b75ad39fe_story.html?utm_term=.22a521a79917.

Federal Emergency Management Agency. 2018. "National Flood Hazard Layer
(nfhl)." Digital database. Department of Homeland Security. Last updated
February 2, 2018. Accessed June 23, 2018. https://www.fema.gov/national
-flood-hazard-layer-nfhl.

Felton, Emmanuel. 2017. "What Happened to All of the New Orleans Teachers
Fired after Katrina?" *Education Week Blog, Teacher Beat*, May 31. Accessed July
3, 2018. http://blogs.edweek.org/edweek/teacherbeat/2017/05/fired_katrina
_teachers.html.

Fessenden, Ford, Robert Gebeloff, Mary Williams Walsh, and Troy Griggs. 2017.
"Water Damage from Hurricane Harvey Extended Far beyond Flood Zones."
New York Times, September 1. Accessed January 16, 2018. https://www.nytimes
.com/interactive/2017/09/01/us/houston-damaged-buildings-in-fema-flood
-zones.html.

First Street Foundation. 2018. "As the Seas Have Been Rising, Home Values Have
Been Sinking." Press Release. *FirstStreet.org*, July 25. Accessed July 26, 2018.
https://assets.floodiq.com/2018/07/ee94ac7b8efe808e9312fa34048e77f6-First
-Street-Foundation-As-the-seas-have-been-rising-home-values-have-been
-sinking.pdf.

Flavelle, Christopher. 2018. "Latest Climate Threat for Coastal Cities: More Rich
People." Bloomberg, April 23. Accessed December 31, 2018. https://www
.bloomberg.com/opinion/articles/2018-12-28/2019-predictions-defiant-trump
-climate-change-and-amazon-hq3.

Gaul, Gilbert M. 2017. "How Rising Seas and Coastal Storms Drowned the U.S.
Flood Insurance Program." *Yale Environment 360*, May 23. Accessed June 26,
2018. https://e360.yale.edu/features/how-rising-seas-and-coastal-storms
-drowned-us-flood-insurance-program.

Harris, Alex. 2018. "The Risk of Sea Level Rise Is Chipping Away at Miami
Home Values, New Research Shows." *Miami Herald*, April 24. Updated April
26, 2018. Accessed June 3, 2018. http://www.miamiherald.com/real-estate
/article209611439.html.

Huber, Dan. 2012. "Fixing a Broken National Flood Insurance Program: Risks
and Potential Reforms," June. Arlington, VA: Center for Climate and Energy
Solutions. Accessed December 28, 2017. https://www.c2es.org/site/assets
/uploads/2012/06/flood-insurance-brief.pdf.

Keenan, Jesse M., Thomas Hill, and Anurag Gumber. 2018. "Climate Gentrification: From Theory to Empiricism in Miami-Dade County, Florida." *Environmental Research Letters* 13 (5): 054001. Accessed June 3, 2018. http://iopscience .iop.org/article/10.1088/1748–9326/aabb32.

Lewis, Maya. 2017. "Climate Change Is Intensifying Gentrification—Here's How." *Everyday Feminism On-Line Magazine*, October 27. Accessed February 1, 2018. https://everydayfeminism.com/2017/10/climate-change-gentrification/.

McAlpine, Steven A., and Jeremy R. Porter. 2018. "Estimating Recent Local Impacts of Sea-Level Rise on Current Real-Estate Losses: A Housing Market Case Study in Miami-Dade, Florida." *Population Research and Policy Review* (June 26): 1–25. Accessed August 2, 2018. https://doi.org/10.1007/s11113–018–9473–5.

Milman, Oliver. 2018. "Flooding from Sea Level Rise Threatens Over 300,000 U.S. Coastal Homes—Study." *Guardian*, June 18. Accessed June 18, 2018. https://www.theguardian.com/environment/2018/jun/17/sea-level-rise -impact-us-coastal-homes-study-climate-change.

Moulite, Jessica. 2017. "Color of Climate: Is Climate Change Gentrifying Miami's Black Neighborhoods?" *The Root*, August 4. Accessed January 31, 2018. https://www.theroot.com/color-of-climate-is-climate-change-gentrifying -miami-s-1797516942.

Natural Resources Defense Council. 2017. "Report: Homeowners Trapped by Repeated Flooding under Troubled Flood Insurance Program." NRDC— Press Release, July 25. Accessed December 31, 2017. http://www.publicnow .com/view/14E73A65BEEC96B7CAAE4965890BCB7F0E6BCAEB?2017–07 –25–19:30:07+01:00-xxx990.

New Jersey Department of Environmental Protection. 2018. "Permit Applications Filed or Acted Upon: Flood Hazard Area." *DEP Bulletin* 42, Issue 12 (June 20): 39 (Shuchter). Accessed January 5, 2019. https://www.nj.gov/dep/bulletin /bu2018_0620.pdf.

Peloso, Margaret E. 2017. *Adapting to Rising Sea Levels: Legal Challenges and Opportunities*. Durham, NC: Carolina Academic Press.

Pinter, Nicholas, Nicholas Santos, and Rui Hui. 2017. "Preliminary Analysis of Hurricane Harvey Flooding in Harris County, Texas." California Water Blog, UC Davis Center for Watershed Sciences, September 1. Accessed January 31, 2018. https://californiawaterblog.com/2017/09/01/preliminary-analysis-of -hurricane-harvey-flooding-in-harris-county-texas/.

Sack, Kevin, and John Schwartz. 2018. "As Storms Keep Coming, FEMA Spends Billions in 'Cycle' of Damage and Repair." *New York Times*, October 8. Accessed October 8, 2018. https://www.nytimes.com/2018/10/08/us/fema-disaster -recovery-climate-change.html.

Satija, Neena, Kiah Collier, and Al Shaw. 2017. "Houston Officials Let Developers Build Inside of Reservoirs. But No One Warned Buyers." *Texas Tribune*,

October 12. Accessed April 15, 2018. https://apps.texastribune.org/harvey-reservoirs/.

Schouten, Cory. 2017. "Climate Gentrification Could Add Value to Elevation." CBS *Money Watch*, December 28. Accessed January 31, 2018. https://www.cbsnews.com/news/climate-gentrification-home-values-rising-sea-level/.

Sladky, Lynn. 2016. "Evicted at Trailer Park 'Paradise,' Residents Seek New Home." Associated Press/*San Diego Union Tribune*, August 31. Accessed April 15, 2017. http://www.sandiegouniontribune.com/sdut-evicted-at-trailer-park-paradise-residents-seek-2016aug31-story.html.

Union of Concerned Scientists. 2018. "U.S. Coastal Property at Risk from Rising Seas." Accessed June 18, 2018. https://ucsusa.maps.arcgis.com/apps/MapSeries/index.html?appid=cf07ebe0a4c9439ab2e7e346656cb239.

U.S. Government Accountability Office. 2010. "National Flood Insurance Program: Continued Actions Needed to Address Financial and Operational Issues." A Testimony before the Subcommittee on Housing and Community Opportunity, Committee on Financial Services, House of Representatives, 11th Congress. Statement of Orice Williams Brown, director, Financial Markets and Community Investment, Federal Emergency Management Agency. GAO-10–631T, April 21. Accessed July 3, 2018. https://www.gao.gov/products/GAO-10–631T.

U.S. Government Accountability Office. 2008. "Flood Insurance: FEMA's Rate-Setting Process Warrants Attention." GAO Report 09–12. U.S. Government Accountability Office, October. Accessed July 3, 2018. https://www.gao.gov/assets/290/283040.html.

Vasilogambros, Matt. 2016. "Taking the High Ground—and Developing It." *Atlantic*, March 6. Accessed April 1, 2018. https://www.theatlantic.com/business/archive/2016/03/taking-the-high-ground-and-developing-it/472326/.

Yeo, Sophie. 2018. "If Americans Keep Ignoring Flood Risk, We Could Face a Housing-Market Crash." New Landscapes, *Pacific Standard: The Social Justice Foundation*, May 21. Accessed June 2, 2018. https://psmag.com/environment/underestimated-flood-risk-could-crash-the-housing-market.

Chapter 11. What You Can Do about Sea Level Rise

Chester, Mitchell. 2018. "Coastal Residents Need to Set Aside Money Now to Cope with Future Flooding—Opinion." *Sun Sentinel*, July 10. Accessed July 24, 2018. http://www.sun-sentinel.com/opinion/fl-op-viewpoint-sea-level-rise-flooding-costs-chester-20180705-story.html.

National Oceanic and Atmospheric Administration. "National Storm Surge Hazard Maps—Version 2." National Hurricane Center. Accessed July 4, 2018. https://www.nhc.noaa.gov/nationalsurge/ and "National Storm Surge Hazard

Maps." http://noaa.maps.arcgis.com/apps/MapSeries/index.html?appid
=d9ed7904dbec441a9c4dd7b277935fad&entry=1.

Sweet, W. V., R. Horton, R. E. Kopp, A. N. LeGrande, and A. Romanou, 2017.
"Sea Level Rise." In *Climate Science Special Report: Fourth National Climate
Assessment, Volume I*, edited by D. J. Wuebbles, D. W. Fahey, K. A. Hibbard,
D. J. Dokken, B. C. Stewart, and T. K. Maycock. Washington, DC: U.S. Global
Change Research Program, 333–63. Accessed December 27, 2018. doi: 10.7930
/J0VM49F2. https://pubs.giss.nasa.gov/abs/sw03000d.html.

· INDEX ·

Page numbers followed by *f* indicate illustrations. Page numbers followed by *t* indicate tables.

farming, 77–78
  effect of salinization on, 77
  effect of salt water intrusion on, 77–78
  impact of sea level rise on, 77–78
  Pamlico and Albemarle sounds, 77–78
  Sacramento Delta, 77
Federal Emergency Management Agency
    (FEMA), xv, 21–22, 24, 92, 99, 105,
    121–23, 125, 127, 129, 137, 142, 153
Federal Flood Insurance. *See* National
    Flood Insurance Program (NFIP)
First Street Foundation, 134
Flavelle, Christopher, 133
Florida Keys, 91, 133, 147
  road raising in, 91
Florida Mobile Home Act, 130
Florida Power and Light (FPL), 82–83
Florida Sun Sentinel, xiii, 1, 65
FM Global, xiv
Fourth National Climate Assessment, xiii
Fugate, Craig, 129

Galveston, Texas, 66–67, 76, 106, 116, 118,
    133
  raising community, 91
  seawall, 66, 66f, 69f
Ganges River Delta, Indian and Bangla-
    deshi refugees, 7
gentrification, 130–34
  paths to gentrification, 131–32
  postdisaster gentrification, 132
Given, Suzan, 85
Gladwin, Hugh, 130
Global Carbon Project, xiii
Government Accounting Office, 122
Grand Isle, Louisiana, 74

Hampton, Virginia, 99
Hampton Roads, Virginia, 88, 132
hard stabilization, 62, 65–69
  cost, 152
Hauer, Mathew:
  refugee numbers, 5, 7, 10, 12

Hawaii, 62, 71, 75–76, 97, 98f
Hawaiians, 24–25
home buyouts, 14, 106, 115, 128
Homeowner Flood Insurance Affordabil-
    ity Act, 125–27
home value loss, 134–35, 135t
hot spots, 9–13, 33, 109
  nuisance flooding, 5
  sea level rise: Inner Banks, South
    Florida, and Mississippi Delta, 9
Houma, Louisiana, 13, 22
Houston, Texas, 37, 39, 85, 91, 97, 104–6,
    114, 118–19, 122, 126, 133
  Hurricane Harvey flooding, 124
  reservoirs: Barker and Addicks, 104,
    120, 124–25
  Ship Channel, 105–6
Hummel, Michelle, 86–87
Hurricane Betsy (1965), 122
Hurricane Bhola (Bangladesh, 1970), 13
Hurricane Dolly (2008), 133
Hurricane Florence (2018), 92
  coal ash spills, 92
Hurricane Floyd (1999), 9, 145
Hurricane Harvey (2017), 5, 37, 39, 85,
    91–92, 104–7, 120, 122–26, 133
Hurricane Ike (2008), 66, 103, 133
Hurricane Irma (2017), 40, 130, 133, 145
Hurricane Katrina (2005), 2, 13, 36, 90, 97,
    104, 108–9, 132–33, 145–46
Hurricane Maria (2017), 1, 38, 39–40, 93
  post-Maria health problems, 39
Hurricane Matthew (2016), 92
  coal ash spills, 92
Hurricane Rita (2005), 13, 145
  escaping Rita, 138f, 145
Hurricane Sandy (2012), 2, 14, 84–85, 97,
    99, 115, 117, 119, 124
hurricanes:
  communication after the storm, 93–94

Ige, David, governor of Hawaii, 97
Ike Dike, 106, 116–17